Public Access Cable Television in the United States and Canada

Gilbert Gillespie

The Praeger Special Studies program—utilizing the most modern and efficient book production techniques and a selective worldwide distribution network—makes available to the academic, government, and business communities significant, timely research in U.S. and international economic, social, and political development.

Public Access Cable Television in the United States and Canada

With an Annotated Bibliography

PRAEGER SPECIAL STUDIES IN U.S. ECONOMIC, SOCIAL, AND POLITICAL ISSUES

Praeger Publishers   New York   Washington   London

Library of Congress Cataloging in Publication Data

Gillespie, Gilbert.
    Public access cable television in the United States
and Canada.

    (Praeger special studies in U.S. economic, social,
and political issues)
    Bibliography: p.
    1. Community antenna television—United States.
2. Television programs, Public service. I. Title.
HE8700.7.C6G54   384.55'47   74-11602
ISBN 0-275-09980-6

PRAEGER PUBLISHERS
111 Fourth Avenue, New York, N.Y. 10003, U.S.A.

Published in the United States of America in 1975
by Praeger Publishers, Inc.

Printed in the United States of America

This book is provided as a reference volume for persons who are involved with or planning to get involved with the production of community media programing. A new idea for developing local communication that deserves serious consideration by community-minded citizens is public access (community) cable television (PACT). In this volume the reader will find a survey of public access history and a discussion of the apparent impact of early experiments with the PACT idea. Someone has stated that the prime value of PACT is the decentralization of the sources of propaganda. This book contains considerable reference to the possible democratizing effects of public access to the media in a time when white-collar crime is on the rise and a widening credibility gap is developing between community leaders and institutional leadership at the city, state, provincial, and national levels. Portapak cable television programing appears to be the medium from which the public access idea generates most profusely. There are some experiments dedicated to developing public access to a combination of the media of television, radio, film, and print, with portapak videotaping being the predominant tool.

If commercial television networks become more responsive to local needs, it might be due, in part, to the new pressure developed by PACT programing. In an interview by Stephani Keer on September 6, 1974, for The Albertan, Don Carroll, director of television for the new Canadian Broadcasting Corporation station in Calgary fervently avowed, "We must feel the pulse of what the city wants and respond and respond and respond and respond." It is hoped that this book will be a useful response to the wants of those who are enthusiastic about participating in the creation and communication of community media organized around cable television.

Thanks is due Cal Downs and Mary E. Curtis for bringing this book into being.

Public Access Cable Television in the United States and Canada

PUBLIC ACCESS
CABLE TELEVISION
IN NORTH AMERICA

As an obsolete definition of the word "discourse,"
Webster lists the term "social familiarity." Beginning in
the late 1960s in the United States and Canada a new means
of social intercourse or discourse, reminiscent of quaint
town-meeting or town-pump rhetoric of New England, has
rapidly evolved in direct response to an ever-intensifying
need for a return to social familiarity in our communities.
This new means of community discourse is labeled public access (community) cable television (PACT) in the United
States and community television (television communautaire)
in Canada. It is a return to decentralized community communication of the vernacular processed in the vernacular.
In this local interaction the community benefits from control of its own affairs as the cybernetic concept of feedback is applied to the major interpersonal problems of the
community. Recently, in a summary of a report prepared
for the Texas Education Agency, Frank Korman of the Center
for Communications Research at the University of Texas at
Austin made these two statements: "[1] The educational
and social impact of cable technology is likely to be
greater than that of any other foreseeable advance in telecommunications technology. [2] The social impact of cable
will alter our basic economic and political institutions
in unforeseen ways."[1] The possibility that cable technology will impose unidentifiable change of considerable impact upon our economic and political institutions in the
near future gives impetus for an examination of the current
state of this techno-evolution. The present examination
will expand from public-access to mass cablecasting in
metropolitan North America.

In 1968 there began to develop in this country and in Canada a tool of unknown power for opening up new channels of dialogue in those communities capable of originating cable television productions. A McLuhanesque interpretation of CATV indicates that "The role of a cable system is to increase the community's awareness of their existing cultural system thereby giving them more control over its development."[2] The problem identified for this study is the examination of the phenomenon of PACT specifically in terms of its state of development and raison d'etre as stated by those individuals, organizations, and agencies most responsible for its development.

In a statement of the Canadian Radio-Television Commission (CRTC), issued February 26, 1971, in preparation for a public hearing on "The Integration of Cable Television in the Canadian Broadcasting System," there was this introduction to a discussion of programing by cable television:

> In its announcements of May 1969 and April 1970, the Commission referred to community programming on cable television placing emphasis on the opportunity for licensees to: "enrich community life by fostering communication among individuals and community groups. In the development of programs of interest to the communities, it is the hope that cable television programmers will be motivated by innovation rather than imitation. Local programmes should be based on access and freedom from the restraint of program schedules which are often less flexible in conventional broadcasting."
>
> The encouragement given by the Commission to the principle of community programming has resulted in many interesting experiments. It is now possible to conclude that community programming is a practical possibility and that it has considerable potential for further development and refinement as part of the Canadian broadcasting system.[3]

Part of the task is to survey those "interesting experiments" that now exist in North America and ask why community programing is thought "a practical possibility . . . that . . . has considerable potential for further development and refinement." Partly, too, it is to conduct a historical study of the issues and policies related to

community application and utilization of PACT. But, most
important, the task is to investigate the role of city
government in the development of PACT in the 150 major
metropolitan areas of the United States and Canada.

"Free access public TV channels have the potential to
revolutionize the communication patterns of service organizations, consumer groups, and political parties, and
could provide an entirely new forum for neighborhood dialogue and artistic expression."[4] This assessment of the
potential impact of public access cable television on our
communities comes from the Center for Analysis of Public
Issues at Princeton. In July 1972 CRTC released an interim
summary of information obtained by its Broadcast Programmes
Branch from a preliminary analysis of 82 questionnaires
returned by Canadian cable television operators. The questionnaire inquired after basic data on local programing in
the United States. No report exists in either country on
the comparative involvement of the major cities in the development of PACT. Don Fabun notes that a theory suggested
by the Canadian philosopher Marshall McLuhan "is that the
effect of a major technological change upon society is to
cause it to become 'numb.'"[5] Both the Canadian and U.S.
federal governments have now brought considerable pressure
to bear on their major cities, urging them to include the
entire population in any plans they have for the development of cable television. This not only involves a major
technological change; there is reason to believe that it
would pose the threat of traumatic sociological change.
Cable television promises to be very big business in the
major markets, a very powerful means of two-way communication partially controlled by the general public. This new
hybrid form of mass communication might come as a shock to
cities where there has been a free-enterprise unidimen-
sional communications system efficiently managed by an
elite corps of professional broadcasters. "In societies
suffering a sudden major technological change," Fabun ob-
serves, "this numbness takes the form of rejection, refusal
to admit the existence of change, ridicule and sometimes
outright opposition."[6] If there is "outright opposition"
on the part of some cities toward the provision of public
access to the cable television medium, the greater body
politic should have some means of gaining information
early so that corrective action can be taken. Fabun continues, "We find nothing in our history with which to form
a comfortable image of ourselves in relation to our electronic environment. Nearly all of our hard-earned past
has, in truth, become a 'bucket of ashes.'"[7] And so the

importance of this study lies in the picture it should offer of the major metropolitan governments of North America
in their individual and collected responses to the decree
or mandate to develop PACT. This picture is given in the
face of their historical referents disintegrating into the
proverbial "bucket of ashes."

A study of PACT is also important at this time for the
following reason proffered by William Stroud of the Department of Communications at the University of Wisconsin, Mil-
waukee:

> The regulatory policy at the federal governmental
> level has just become effective in spring (April)
> 1972. The implementing actions by state or munic-
> ipal authorities in many places are still in de-
> velopment. This situation poses for the public
> and the country at large, and particularly for
> scholars of the mass communication media, an op-
> portunity to play some part in shaping the future
> of the cable communications industry.[8]

The historical sketch of PACT presented in this study
should support the view that new and extensive developments
that are afoot in the field of telecommunications can work
considerable influence upon the quality of human life.
Federal governments are offering a specific element of this
new sociological force as a sort of servomechanism for controlling the main power holders that tend to dominate communities when there is a scarcity of the channels of intercommunication necessary to the development of innumerable
very small groups of power. Certain cable operators are
required by law to provide free public access to a television channel, but no individual or community is bound
by law to accept the challenge of developing programing
for the free channel. A tremendous job of promotion is
yet to be accomplished. Several communities have taken
up the challenge of developing programing for the free
channels without the pressure of governmental urging that
now prevails. Many communities will, in due time, come
under governmental pressure for a free channel as channels
become available with the introduction of new cable systems
in the larger markets, or with the expiration of the grandfather clause, which has protected established franchise
holders. In a landmark policy statement handed down in
July 1971 CRTC announced: "Cable television systems should
be encouraged to provide access to a channel for community
expression and information."[9] The U.S. Federal Communica-

tions Commission (FCC) made the access channel requirement applicable "to all cable television systems which commence[d] operations in a major television market [top-100] after March 30, 1972"; the grandfather clause applies to systems already in operation prior to March 31, 1972, which are given until March 31, 1977, to comply.[10]

It follows that any community involved with public access might be searching for answers to a number of questions. For example:

- Why PACT for us?
- What is the philosophy behind it?
- What are some means of financing a public access program in our community?
- Is consultant assistance available?
- How do we organize a PACT board or committee that will generate aggressive and imaginative production units?
- What kinds of productions are being done in other communities, for what purpose, and with what sort of community involvement?
- What are the current state and future prospects of the technology of public access?
- How do we develop a viewership?
- Who would our viewers be, and how important is it to attempt to attract a large number of viewers?

By surveying the literature and inquiring with the coterie of pioneers who have been involved with public access development it should be possible to prepare a scholarly document that would have substantial informational value, the like of which has not been available for reference to date. In May 1972 Mayor Charles Wheeler of Kansas City, Missouri, began selection of a seven-man committee for the task of gathering information and making recommendations concerning the type of cable television franchise that should be granted in the city within the year. The ownership decision, the conditions of the franchise, and the public access requirement were very much bound together. One member of that citizens committee, Andrew McCanse, assistant dean of the University of Missouri at Kansas City School of Medicine, indicated to this writer that the city government had not allocated any funds for a consultant's study and that committee members invariably had a very limited amount of time to devote to personal research of the complex ramifications of metropolitan cable television planning. McCanse was of the opinion that the type of study suggested would be of some help in providing insight into one consideration of the committee's general assignment: the vital consideration of public access (community) cable television.[11] The final report of the Kansas City study stated that the community exerted insufficient pressure for cable television on the city government to warrant the award of a franchise. There is no reason to believe that conservative city governments will find the money or the courage in the foreseeable future to risk the revolutionary technology of cable television with its social side effects. An important and timely opportunity for the community was not acted upon, and the local television broadcasters continue to bask in the status quo.

In the United States there are four designations of cable television access channels: public, educational, local government, and leased. All cable systems in this country located "wholly or partially within a top-100 television market which began or begin operations on or after March 31, 1972" must provide those four categories of access channels. The FCC has recognized that this requirement could be an undue burden on some smaller cable operators within the top-100 markets and, therefore, entertains requests for waivers.[12]

In a 1971 brochure designed to promote "TV by the people, for the people" in New York City, Tele-PrompTer Manhattan CATV Corporation's Henry D. Pearson provides a definition of public access cable television as "a whole new concept in television" that affords the opportunity for "any groups or individuals of any belief, purpose, or persuasion, to demonstrate their talents." If the individuals or groups do not have their own equipment, Tele-PrompTer Cable TV, like other public access cable operators, will provide free of charge at least one television camera, a studio, and the services of a director. Tele-PrompTer and other operators make portable equipment available "to cover events in the community, like block parties, park openings and church functions." "Even better," suggests Pearson, "produce the show entirely on your own and bring your film or videotape to TelePrompTer. We'll put you on TV."

The National Cable Television Association lists six general FCC requirements related to the public access channel that summarize this new opportunity in human communication:

1. Cable systems obligated to provide public access must make available at least one noncommercial channel. Under specified conditions of continuous heavy volume of

usage an additional channel must be made available within a period of six months.

2. At least one channel must always be available for public access.

3. Such production equipment in good repair as is necessary to produce a simply staged public access program must always be available.

4. Cable operators may levee a modest charge for production costs of live studio programs that exceed five minutes.

5. Cable operators are required to establish for public inspection and filing with the FCC within 90 days certain operating rules to be applied to the access channel.

6. With the following exceptions cable operators must refrain from exercising any control over program content (these rules have been set by the FCC as being essential to guidance of the cable and the public access producers in their working relationships):

a. Access, whether by individuals or groups, must be maintained on a first-come, nondiscriminatory basis. "The guiding principle," insofar as scheduling is concerned, "should be to give access to the greatest number of users."

b. Prohibitions designated by the FCC for which the operators have some responsibility are copyright clearances, presentation of obscene and indecent matter, lottery information, and commercial promotion of products, services, or candidates for public office. Also, operators must keep a record for a minimum of two years of all names and addresses of persons or groups requesting time.

c. Just as cable operators are expected to provide minimum facilities, equipment, and technical assistance, in return they "may require reasonable technical standards in software submitted by a user."

d. The NCTA encourages cable operators to form "broad-based democratically constituted groups to advise and assist in policy development, promotion and funding of public access."[13]

In Vancouver, Canada, public access is taking on a slightly different definition as the result of experimentation with a media mix. Chris Pinney of Metro Media writes about this development in their July 1972 newsletter:

Our first year has seen a number of changes in our conception of ourselves as an organization and the role we play in Vancouver's development. Most importantly we have realized that to expand citizen communication power we have to be more than just a service organization.

To effectively develop media as a useful tool in building communities we have found it necessary to become an active initiator of new projects and to maintain a strong animating role in dealing with both institutions and government as well as the general public. We have changed over the year from using one medium--VTR, to providing a production service for groups who wish access to cable TV, to working to coordinate a variety of media and communication networks that will allow people to collectively define the kind of communities they wish to live in.

Keeping our original goals in mind we have developed into a community service and animation organization that meets communication needs through a number of related activities:

1. training and animation
2. community journalism
3. information co-ordination and dissemination
4. development of communication networks
5. research and exploration[14]

Since September 10, 1971, Metro Media has had a regular two-hour time slot on Vancouver Cablevision's Channel 10 Friday evenings. Most of this time has been devoted to cablecasting tapes produced by community groups in the greater Vancouver market of 235,000 subscribers.[15]

In essence, the public-access (community) animator is attempting to reconstruct miniatures of McLuhan's "global village"--electronic villages. A return to the intimate communication of yesteryear's New England town is the objective. In a speech delivered at the June 26, 1971, Urban CATV Workshop in Washington, D.C., Commissioner Nicholas Johnson quoted these lines from "An Evening's Journey to Conway, Massachusetts," by Archibald MacLeish:

I'll tell you what's a town. It's a meeting of minds. And how do you get a meeting of minds? Meeting of men. And how do men meet? By the ways . . . Roads! That's what a town is--men going back and forth on their occasions. Pass-

ing each other. And not always passing--pausing
sometimes--speaking--palaver. That's what a town
is . . . a meeting--a meeting of men--minds.

"So it is with cable's potential," affirmed Commissioner
Johnson. "Perhaps as never before has technology given us
the opportunity for a new 'meeting of men, meeting of minds'
in our unworkable inner cities."[16] In the "Telecommunications: One World-Mind" issue of Kaiser News there is this
statement: "If there is any direction to evolution at all,
it appears to be to create more channels for communication;
atom to atom, molecule to molecule, part to part."[17] The
same holds true for PACT.

For the sake of brevity the acronym PACT is used in
place of the term public access (community) cable television throughout the remainder of this study. PACT, it
should be carefully noted, is further defined as community
cable television programing produced and controlled by a
committee of community representatives (a community board),
as opposed to programing originated locally but controlled
by the management of the cable system(s).

NEGENTROPY AND ENTROPY, OR
ARRANGEMENT AND DERANGEMENT

Entropy . . . serves two closely related purposes.
First, it is what Eddington called time's arrow,
that is, a pointer of the natural processes. Sec-
ondly, it reveals to us quantitatively the statis-
tical structure of internal motions very much as
information theory does that of our message ensem-
ble. Moreover, it does so in an analogous way.
For we take entropy as the logarithm of thermodynamics probability of a macrostate just as we
measure information of a message by the logarithm
of the probability of its occurrence. The reason
in either case is to secure additivity. . . .
The logarithmic form enabled us to ensure that
the information content of any complex of messages is the sum of its individual components.
For exactly the same reason we measure the entropy of a body in any macrostate by the logarithm of the number of its corresponding micro-
states, that is, its thermodynamic probability.[18]

Certainly there are scholars who will argue that the information content of any complex of messages is greater than

the sum of its individual components. In this definition
of entropy developed by Jagsit Singh the relationship between entropy and information is distinctly drawn:

> Obviously, information and entropy are two
> sides of the same coin in that internal order or
> organization, and therefore greater knowledge or
> information of the system's internal make-up,
> goes hand in hand with low thermodynamic proba-
> bility or rather its logarithm, which we have
> chosen to call entropy. Since in any given sys-
> tem, the greater the number of microscopic
> states corresponding to any assigned macrostate,
> the greater will be its entropy, it follows that
> entropy is a measure of our ignorance of its ul-
> tramicroscopic structure. In other words, en-
> tropy is information in reverse gear, for which
> reason L. Brillouin coined the term "negentropy"
> by contracting the phrase "negative of entropy."[19]

The striking correlation between quantity of arrangement or derangement of information and negentropy and entropy was discovered by L. Szilard in the process of his
wrestling with the theoretical demon created by J. C. Max-
well's Theory of Heat in 1871.

CHALLENGE FOR CHANGE/SOCIETE
NOUVELLE

Challenge for Change/Societe Nouvelle is an experimental program established by the government of Canada as
a participation between the National Film Board of Canada
and certain federal government departments and agencies.
They now comprise agriculture; central mortgage and housing corporation, national health and welfare; Indian affairs and northern development; labor; and regional economic expansion and secretary of state/citizenship. The
program is responsible directly to the secretary of state,
via the Privy Council office. Challenge for Change/So-
ciete Nouvelle is administered by a committee made up of
six film board representatives and one from each supporting
department. The total budget established for the two programs for 1972-73 was $1.4 million.

THE PURPOSE OF STUDYING THE
DATA GATHERED

It has been suggested that metropolitan government in
North America is faced with a brand new problem of unknown
but profuse dimension. Along with all of the other problems with which they must wrestle, civic bodies must become
thoroughly involved with the consideration of design and
control of a wired city. For the first time there is an
obligation to involve both individual private citizens of
the most humble stature and community communications committees in planning the design and future control of an
all-pervasive and revolutionary factor of city life.
There is now an obligation on the part of the major city
governments of Canada and the United States to maintain
and invite a defined share of access for individuals and
citizens' groups to the proliferating channels of cabled
communication in the city. The city fathers must now nur-
ture and eventually react, if they are not already doing
so, to many new decentralized sources of local propaganda.
This study is designed, therefore, to gain a general impression of how governments of those major cities surveyed
are responding to their new responsibility of facilitating
the wiring of their cities. A specific impression is
sought of how they are dealing with PACT. In addition, an
extensive survey of the literature is reported for the
purpose of providing a general setting in which to place
the findings of the city government survey.
     Concluding the survey of the literature segment of the
study is a summary of those conditions that seem to have
been most important to the success or failure of the pioneer PACT experiments. Through ongoing evaluation of the
changing elements of the PACT idea a practical model, relatively free of corruption, should be maintained.

THE IMPORTANCE OF PUBLIC
ACCESS CABLE TELEVISION

In a speech delivered several years ago to the Execu-
tive Club in Chicago, Arthur R. Murphy, Jr., president of
McCall Corporation, discussed a statement by Marshall
McLuhan:

   "The personal and social consequences of any medium . . . result from the new scale that is introduced into our affairs by each extension of

ourselves, or by any new technology." McLuhan
seems to be saying here . . . that all media have
effects on the human organism that are independent
of the message being carried. The long-term effects of "sense isolation," "extensions of man,"
and "media hot and cold" are much more important
to man's development, than are individual messages. Thus, in this historical sense, the medium is the (important) message.[20]

Philosophically, then, what is the importance of PACT to
the development of man? What are the basic assumptions,
objectives, and processes involved in this new concept of
"village" dialogue? Major dynamic problems and their
causes inherent in the development of public access should
be revealed in examining the processes.

It is important, of course, that there be some logic
to this new experiment with the media. But how does one
determine whether logic exists in this experiment? Margaret
J. Osler points out that when "we cannot know the truth of
a proposition either by intuition or by demonstration, we
can nevertheless judge it to be true or false. . . . Although we judge in the absence of certainty," asserts Osler, "we can nevertheless have good grounds on which to
base the degree of assent we should grant to less-than-
certain propositions; such an evaluation is made on the
basis of probability."[21] Osler made a direct appropriation
of the wisdom of Locke with a quotation of his analysis of
those logical relationships that contribute to human insight into the science of material substances:

As demonstration is the showing the agreement or
disagreement of two ideas, by the intervention of
one or more proofs, which have a constant, immuta-
ble, and visible connexion one with another; so
probability is nothing but the appearance of such
an agreement or disagreement, by the intervention
of proofs, whose connexion is not constant and im-
mutable, or at least is not perceived to be so,
and is enough to induce the mind to judge the
proposition to be true or false, rather than the
contrary.[22]

As has been previously suggested, the "excuse" for
reacting or responding in a particular way to observed
logical relationships in new combinations of material substances is the "stuff" philosophy is made of. Upon exam-

ining the thread of philosophy that appears to weave
through the rationale for existence provided by the several
public access cable producers, one might be reminded of
the new order of reality created by the "ought" found in
the writings of the cognitive psychologist Fritz Heider:
"The supra-individual reality of value and ought is a phenomenon of a new order of complication. It is an emergent
social phenomenon which has to do with the exclusion of
individual wants and likes, and which brings with it new
constant functions, new perceptions and new possibilities
of action." Heider goes on to explain, "the ought is not
merely a feeling, some esoteric quality that can be
glimpsed by the phenomenologist in a happy moment. It in-
fluences real events."[23] To clarify further the concept
of the influential ought, Heider turned to the philosophical writings of George Mead and his concept of the "generalized other":

> The attitude of the generalized other represents
> the attitude of the whole family. The individ-
> ual is confronted with the attitude of the gener-
> alized other, attitudes that have greater objectiv-
> ity than his own personal wants and attitudes.
> The individual can put himself in the place of the
> generalized other, and that assures the existence
> of a universe of discourse. Thus, the individual's
> conduct is guided by principles, "and a person who
> has such an organized group of responses is a man
> whom we say has character, in the moral sense."
> Mead illustrates this taking over of the attitude
> of the generalized other by referring to the no-
> tion of property. Property is that "which the
> individual can control himself--nobody else can
> control." If the attitudes referring to property
> are established, and new ways of functioning are
> prescribed for the individual. If p owes some-
> thing to o, then the objective order wishes p to
> give it to o.[24]

If the concept of the generalized other is applied to
the mass media (television, in particular), considerable
concern should develop as to the peculiar "universe of
discourse" that does or does not result. This concern
would seem to be manifest in the McLuhan rhetoric that has
served to provoke contemporary thinkers to analyze the
sociological impact of mass telecommunication. If the
"personal wants and attitudes" of the broadcasters are

fixated on the righteous free enterprise path of responding only to the personal wants and attitudes of the lowest
common denominator in audience viewership, then "all the
different rights and duties in regard to [that public]
property," the airwaves, supposedly guaranteed by the FCC,
remains a mythical supposition. Public access advocates
might well rest their case on this argument alone and
then go on to assert that if network broadcasters do not
respresent the attitudes of "the whole community," there
does exist a potential within the telecommunication medium
for a much more comprehensive group of responses from men
who we can truly say "have character in the moral sense."
A look into the opportunities offered by PACT points to
the possibility of serving the personal wants and attitudes
not only of the bourgeoisie, as prescribed by the rating
charts, but of everyone, including the radical ends of the
political, economic, and social spectrum in every wired
village. Is there really any moral and social necessity
for man to get involved with all of those others of the
generalized other? It would appear that it has always been
more profitable to neglect totally a substantial minority
of those generalized others (at least in the materialistic
sense), even in the management of that most precious of
all public properties, the airwaves of mass communication.
     "What has disappeared--for better or for worse--are
great and noble reasons for living," stated the noted
professor of communication Lee Thayer in a paper prepared
for the First World Conference on Social Communication for
Development in March 1970. In his closing statement Thayer
inquired, "Can a man achieve a sense of human dignity out
of being recognized by the National Broadcasting Company
as 'an average television viewer'? If a society can exist
without these fundamental values, would it be the kind of
society, which is for people?" Such a society, in Thayer's
view, is one that is dedicated to the reduction of entropy
in our monolithic broadcasting structure through increased,
not decreased, decentralization and specialization of the
media:

     What is needed in the proliferation of subgroups
     and subcultures within a state are structures
     which are interdependent in ways that require in-
     tercommunication. A neighborhood of home owners,
     for example, will take to each other, and may be-
     gin to talk to various community leaders and gov-
     ernment officials, about such things as taxes.
     If we are to have enriching, viable societies,

we need to have less, not more, canned information.[25]

Thayer's approach to the problem of entropy is to replace those people involved in the mass production and dissemination of messages who are either incompetent or inadequate, or both. He makes no mention of the possibility of public access cable television being a viable approach to reduction of entropy in the intercommunication face of a community.[26]

In reference to the general impact of a wired universe William T. Knox asked:

> How is the transition between the present state and the future system to be managed? Noticeably lacking in the issues presented by the various government and private interests so far is much concern about the social, psychological and cultural consequences of a nationwide broad-band communications network capable of handling all of an individual's communication needs. Yet this may be the area where the most unforeseen events will materialize.[27]

In the background of the problem, then, appears the instruction to inquire with those responsible for development of PACT to determine how their philosophy relates to a technological breakthrough that has the potential for "handling all of an individual's communication needs." In the spirit of Herbert Parker's analysis of the term "scholarship," a look into the background of the problem points to a definite need for a scholarly integration of knowledge concerning the development of public access:

> "Research" is an unfortunate term. I prefer to think of what scholars write as "scholarship." If they have learned their discipline, the distinction between scholarship and other forms of knowledge becomes trivial.
> What is "scholarship"? It is a learned monograph, complete with footnotes. It is a proposed solution to a pressing problem. It is the discovery that x really is a problem. It is a "popular" look, synthesizing for the intelligent general reader knowledge that has already become known to the specialist. It is all of these.
> . . . The ideal scholar has managed to write not merely for his fellow-specialist, but for the general educated and intelligent public.

15

. . . The man who tries to integrate knowledge
is much the superior of one who froments it.[28]

There has been time for very few specialists to develop in
the field of PACT.

NOTES

1. Frank Korman, "Innovations in Telecommunications
Technology: A Look Ahead," Educational Broadcasting Review 6 (October 1972): 328.
2. Michael H. Molenda, "The Educational Implications
of Cable Television (CATV) and Video Cassettes: An Annotated Bibliography," Audiovisual Instruction 17 (April
1972): 42.
3. CRTC, The Integration of Cable Television in the
Canadian Broadcasting System, a statement, February 26,
1971, in preparation for the public hearing held April 26,
1971, in Montreal, p. 26.
4. Center for the Analysis of Public Issues, Public
Issues, supplement No. 1 (July 1971), p. 1.
5. Don Fabun, The Dynamics of Change (Englewood
Cliffs, N.J.: Prentice-Hall, 1967), p. 26.
6. Ibid., p. 26.
7. Ibid., p. 27.
8. William Stroud, Selected Bibliography on Telecommunications (Cable Systems) (Madison: Wisconsin Library
Association, 1972), p. i.
9. CRTC, Canadian Broadcasting: "A Single System,"
a report (July 16, 1971), p. 13.
10. Federal Register 37 (July 14, 1972): 13859.
11. Conversation with Andrew McCanse, September 2,
1972, in Kansas City, Missouri.
12. The National Cable Television Association, Guidelines for Access: A Report by NCTA (August 1972), p. 1.
13. Ibid., pp. 2-5.
14. Chris Pinney, Metro Media Print-Out 1 (July 1972):
1.
15. Ibid., p. 7.
16. Nicholas Johnson, speech to Urban CATV Workshop,
Washington, D.C., June 26, 1971, Public Notice, Federal
Communications Commission, p. 13.
17. Kaiser Corporation, Telecommunications: One
World-Mind, no. 6 of "The Markets of Change" series (1971),
p. 40.

18. Jagsit Singh, <u>Great Ideas in Information Theory, Language and Cybernetics</u> (New York: Dover, 1966), p. 75.

19. Ibid., p. 76.

20. Arthur R. Murphy, Jr., "Communications: Mass Without Meaning," <u>Vital Speeches of the Day</u> 33 (July 1, 1967): 573-76.

21. Margaret J. Osler, "Locke and the Changing Ideal of Scientific Knowledge," <u>Journal of the History of Ideas</u> 31 (January-March 1970): 39.

22. John Locke, <u>An Essay Concerning Human Understanding</u>, Vol. II, ed. A. C. Fraser (New York: Dover Publications, 1959), p. 363.

23. Fritz Heider, <u>The Psychology of Interpersonal Relations</u> (New York: Wiley, 1958), p. 228.

24. Ibid.

25. Lee Thayer, "On Human Communication and Social Development," <u>Economies et societies: La communication II</u>, Vol. V (Geneva, Switzerland: Librairie Droz, 1971), p. 1628.

26. Ibid., p. 1631.

27. William T. Knox, "Cable Television," <u>Scientific American</u> 225 (October 1971): 22-29.

28. Herbert L. Parker, "Piling Higher and Deeper; The Shame of the Ph.D.," <u>Change</u> 2 (November-December 1970): 53.

2

THE BACKGROUND
OF PUBLIC ACCESS
CABLE TELEVISION

THE HISTORY OF CABLE TELEVISION

The history of PACT must begin with some reference to the historical development of cable television in general. Forest H. Belt offers a synopsis that seems to suit the immediate need of this study:

> Cable television is an enigma. The further it goes, the harder it is to tell just where it is going. When it was simply community antenna TV (CATV), it provided the simple, basic service of bringing TV reception to those areas that broadcast stations could not reach. But 22 years have brought about changes. Purposes and outlook have changed--not to mention the name.[1]

There might be some value in expanding Belt's term "outlook" to read philosophical outlook. To complete Belt's historical sketch:

> Today's cable TV is a system for delivering many channels of TV to home receivers by wire (coaxial cable). Some channels deliver nearby station programs that you would receive anyway. Others carry signals imported from distant stations, picked up by powerful antennas and relayed by microwave. At least one channel can carry programs generated by the cable owner (often including commercial advertising). This is no longer a simple multi-customer service. It is cablecasting. The term "CATV" belongs to a bygone era.[2]

Belt goes on to suggest that the term "cablecasting" might soon belong to a bygone era also: "Just around the corner is two-way cable TV. It will open a whole new era. With it you get into true communication by cable. The term to describe such a system might be cablecom."[3] And inextricably bound up with the historical development of cablecasting and "cablecom" is PACT. The War on Poverty conducted by the federal government of Canada spawned the experimentation with community film communication that eventually resulted in community cable communication. The new portable video technology offered a viable means of opening up a multitude of channels of communication that could serve to effect social change.

As Arthur C. Clarke has stated, "What we are building now is the nervous system of mankind, which will link together the whole human race, for better or worse, in a unity which no earlier age could have imagined."[4] Charles Tate offers this comment:

> In this age of technology, cable television is a super technology--a synthesis of radio wave electronics and computer technology. It's a powerful illustration of Alvin Toffler's theory that technology feeds on itself and that the elapsed time between technological innovations has been so drastically reduced that man is already living in the future.[5]

In September 1970 Hubert J. Schlafly made the "blue sky" prognostication that laser technology "might permit transmission of almost unlimited numbers of TV channels without interference and without costly demodulation equipment."[6] Schlafly concludes his report with these words of encouragement, "A wonderful communications opportunity has been created, almost by accident. Let us use it to good advantage."[7]

An appliance dealer in Lansford, Pennsylvania, is usually credited with touching off the CATV accident in 1949 or 1950. However, E. Stratford Smith provides the documented story of the actual origin:

> There is some good-natured dispute among pioneers in the industry concerning who built the first cable system. The writer, who represents two of the East Coast contenders professionally, will not jeopardize his rapport with either of them by taking sides. However, L. E. "Ed" Par-

sons (who is no longer in CATV) has the best documented claim and is generally credited with having constructed the first cable system in the country at Astoria, Ore., in 1949.

Parsons, then the operator of Radio Station KAST in Astoria, reportedly responding to the challenge of his wife who wanted "pictures with [her] radio," went "searching all over Clatsop County, Ore." with signal-survey equipment for the signals of television station KRSC-TV, 125 miles away in Seattle, Wash. He finally settled, oddly enough, for an antenna site on the roof of the two-story John Jacob Astor Hotel in the center of downtown Astoria where he was living in a top-floor apartment; he discovered a fairly reliable but not very strong signal. After once developing a "watchable" picture in his apartment, Parsons developed a three-tube sending unit and extended service to the hotel lobby and then to a nearby music store. Soon other locations including residences were attached and the service developed thereafter essentially as a cooperative. An installation charge of $100 was collected from each person, who was then regarded as owning the cable facility required to extend service to him from the immediately preceding connection. Parsons also sold television sets to those persons whose residences were connected to the cable.[8]

Very shortly after Parsons' system became operative, appliance dealers in Lansford, Mahanoy City, and Pottsville, Pennsylvania, independently developed their own wired extensions of distant television signals.[9] A report prepared by the League of Oregon Cities begins with the statement: "The first cable television distribution system in this country functioned experimentally in Astoria in 1949."[10]

In 1949 Rediffusion of England began distributing radio signals in Montreal and by 1952 began cablecasting television signals. In the same year CATV was launched in London, Ontario, by E. R. Jarmain, a dry cleaning operator who experimented with electronics as a hobby. Jarmain had no development capital for his 15-household system, and finally in 1959 he entered a partnership with Famous Players Corporation of Canada. By 1970 the new partnership, London TV Cable Service Limited, attained a dramatic 82 percent penetration of the London cable market. Because of the FCC's protection of the broadcast television

stations in the major markets through a six-year freeze,
cable television has developed more rapidly in Canada
than in the United States. By 1970 penetration by the
largest U.S. system in San Diego was 20 percent, while in
the urban centers of Vancouver and Victoria, British Columbia, penetration had reached 65 percent.[11] As of the May
15, 1972, edition of Broadcasting there were about 2,750
operating cable systems in the United States. Another
1,950 systems had been approved for construction, and
2,900 applications were pending before local governments.
It was also reported that over 400 systems had the capability of originating programs, and nearly 300 did so on
a regularly scheduled basis--an average of 16 hours a
week. There was no mention of public access programing.
Also interesting was the fact that about 42 percent of the
cable industry was owned by other communication interests:
30 percent by broadcasters; 7 percent by newspaper publishers; 5 percent by telephone companies. Subscription
revenue in 1971 reached $360 million. In an April 1972
news release The National Cable Television Association
announced that industry leaders estimated that, "assuming
reasonable regulation, by 1980 the CATV industry will
serve 25-30 million homes via nearly 5,000 systems, and
will have annual revenues of over $2 million and a net
worth of $5 billion."[12]

THE NATIONAL FILM BOARD EXPERIMENTS
WITH SOCIAL CHANGE

Within this multibillion-dollar accident called cable
television there rests that promising facet, PACT, which
is also a derivative of serendipity. The National Film
Board's Challenge for Change filmmaker Fernand Dansereau
has related how "by accident" the concept of PACT leaped
ahead at one point in its evolution.
Several years ago Challenge for Change filmed a documentary of a poor neighborhood in Montreal. When this
film, Septembre 5 at Saint-Henri, went into distribution,
an extremely negative reaction developed on the part of
the documented subjects. They expressed feelings of deep
hurt as a result of the public revelation of their mean
existence. Feelings of shame were created by neighbors
who cruelly ridiculed and scorned them, having viewed or
heard about the film. One family was so stung by the
abuse that they resorted to removing their children from
the local school. "Because of the severity of these re-

21

percussions," wrote Dansereau (who, incidentally, was not the producer of the film), "a feeling of deep remorse has remained with me in spite of our undeniable good will."[13]

The National Film Board assigned Dansereau to a similar project in December 1966 in which he documented individuals and institutions in the small Quebec town of Saint-Jerome as they reacted to a period of rapid sociological change imposed in part by governmental decree. Dansereau was moved to search for a more human approach to filmic documentation. His new strategy began with an invitation to all of the anticipated subjects to censor any objectionable portions of the film in which they appeared (politicans were excepted from this censor role). One result of this experiment was a much stronger identification of the film makers (Dansereau, his cameraman, and his sound man) with the subjects of their filming and their community. The Film Board cineaste provides this description of the kind of change that appeared to occur with the subjects of the film:

> At different stages of filming, we invited our principal participants to screen rushes with us. We continued to do so at different stages of editing. This, of course, taught us a great deal and changed us considerably. But it also changed the people. After seeing themselves in the footage as in a sort of mirror, yet with all the security that surrounds an event that has been lived, is known, and is past, they were free to criticize themselves and to decide to change themselves if they felt the need. I am speaking, of course, not from a film-making point of view, but from the point of view of the very being of these persons. In other words, the screenings could, in certain cases, exercise an effect of collective therapy.[14]

"I must add right away, however," he said, "that we had not planned anything in this direction. It occurred, when it did occur, as a sort of accident."[15] Dansereau also observed that subject participation in the total filmic process contributed to the development of a common bond of acceptance of the final product. When the Saint-Jerome chamber of commerce later asked that the film be restricted to certain audiences, the principal community organizations that had collaborated on the production rose up in defense of the film. When distributed to other com-

munities in Quebec with similar French Canadian characteristics, the film seemed to have an audience impact of considerable significance. The considerable discussion it
evoked probably produced some social action, although
Dansereau notes there were no social scientists assigned
to document this possibility. The unique relationship of
the film makers and their character collaborators also
caused Dansereau to offer the smaller audiences a series
of satellite films that could be used to involve the viewer
in developing "the first perception of the reality that
the main film had offered."[15]

David Gee, secretary of the Interdepartmental Committee of the Challenge for Change program, had these reflec-
tions on the purposes and achievements of the film board's
relatively new social arm:

> The purpose of the Challenge for Change pro-
> gram is to create in Canadians an awareness of
> the need for change in order that they may achieve
> a better quality of life. The film medium permits
> people not only to become aware of problems fac-
> ing them in their society, but of government pro-
> grams that can offer real solutions to these prob-
> lems.
>
> Challenge for Change also has effects upon
> the 21 government departments and agencies and
> the National Film Board, which together consti-
> tute the Interdepartmental Committee.
>
> [. . .]
>
> Wrestling with the ideas that come out of
> this milieu, NFB has put together 34 films to
> date. Some have pleased some committee members,
> and some have not. Creation is seldom a painless
> effort. But what they have done is stimulate
> thousands of persons across Canada into seeing,
> perhaps for the first time, an objective and of-
> ten stark picture of their lives, and what they
> can do to improve them. A product such as this
> is worth all the turmoil and anxiety that it pro-
> duces.[16]

Jagsit Singh begins Great Ideas in Information Theory,
Language and Cybernetics with this inquiry into the origin
of natural intelligence:

Although the study of human speech and language is not the most direct approach to how the brain functions, there is perhaps a grain of truth in Fournie's surmise that human speech may well be a window through which the physiologist can observe cerebral life. He may have been led to it by Paul Broca's discovery of the fact that aphasia or loss of speech is caused by destruction of a certain relatively small area of the cortex in the dominant hemisphere of man. This is not to say that speech is located there in the way the United States gold reserves are hoarded at Fort Knox, but that this small area is used as an essential part of a functional mechanism employed while the individual speaks, writes, reads, or listens to others who speak. For despite the damage to the area in question a man can still think and carry out either forms of voluntary activity even if the speech mechanism is paralyzed. In other words, the human engine continues to run although it has ceased to whistle.[17]

Such is the condition that prevails in many of our communities today. One such community that fell victim to "aphasia" was Fogo Island, Newfoundland, where the PACT concept was originally actualized on film in English-speaking Canada. As told by Sandra Gwyn:

That story starts in the summer of 1967, Expo Summer, when the Extension Service and the Challenge for Change unit of the National Film Board, two fledgling social innovators full of promise and untried potential, converged on the Island of Fogo to introduce a new concept of community development. First established in 1959, the Extension Service had newly expanded its mandate to become directly involved in social change in Newfoundland. The Challenge for Change program had just been set up, in the wake of the War on Poverty, to find new ways of using film to provoke such change.

Fogo Island was chosen for the experiment because it represented, in microcosm, most of the basic problems of rural Newfoundland. Lying ten miles off the northeast coast, Fogo, like Newfoundland itself, was isolated from without. Again, like Newfoundland, as a whole, Fogo was

isolated within itself. Fewer than five thousand
people lived on the Island, but they lived sealed
off from one another by religion and background
in ten tiny settlements.[18]

Frontier individualism that has come to haunt so much
of North America as a force of brutal indifference to the
decay of once proud capital cities manifested its parochial
form on Fogo Island. Gwyn describes this parallel of depressed human spirits dwelling miserably incommunicado
with the greater community of man:

> Communication between communities was so poor
> that no one on Fogo ever spoke of it as their
> home, instead they would name their own outport:
> Joe Batt's Arm, Seldom Come By, Stay Harbour, or
> Tilting. In each village, each denomination had
> its own one or two room denominational school.
> A child could grow up without visiting a settle-
> ment five miles away. The decline of the inshore
> fishery had placed sixty percent of the popula-
> tion on welfare. Only one community had local
> government; there were no active unions or pro-
> ducer co-operatives. The people were trapped in
> a cycle of isolation and poverty from which they
> lacked the knowledge and confidence to escape.
> A single future remained for them: resettlement
> at the government's discretion.[19]

This condition has its direct counterparts, certainly, in
Appalachia, in the innumerable ghettos, in the sharecrop-
ping South, and in the migratory worker camps of the fu-
tureless farmers of the United States where the Great De-
pression goes on and on. Gwyn, the chronicler of the
Memorial University Seminar, continues:

> The Memorial-Film Board team was headed by Colin
> Low, a senior Board producer, best known perhaps
> for Corral, Universe and as co-director of Laby-
> rynth. "In film I'd never been much of an action
> man," he told the Seminar. "I'd always been more
> interested in using it for poetic document. But
> in Newfoundland I discovered something very dif-
> ferent." He became the D. W. Griffith of the
> Fogo Process.[20]

It has been noted how serendipity seems to have played
a major role in the origin and growth of cable television.
But it was not really an accident, Gwyn writes, that the
utilization of film as a catalyst for social change, as an
instrument of consensus, first occurred in Canada:

> From Harold Innis, Marshall McLuhan, the innova-
> tions of Expo and Labyrynth, to a world lead in
> cable television (because of course, we all like
> to watch American programs), communications have
> always been uniquely part of the Canadian experi-
> ence. As Neil Compton has written, "The charac-
> teristic virtues of our native tradition--the rec-
> ognition of human limitation, the awareness of am-
> biguity and the urge to communicate--are those
> which the age seems to demand." In a sense, us-
> ing cinema as catalyst amounts to a marriage be-
> tween "the medium is the message" and the National
> Film Board's proud old tradition of social docu-
> mentary. By 1972, the Fogo Process had been
> adapted for use all over North America; the NFB's
> Challenge for Change and Societe Nouvelle units
> seeded community communications groups all over
> the country, who produced local programs for CATV
> Systems or VTR tapes for themselves.[21]

In reviewing the history of the National Film Board
we meet with two film makers of outstanding proportions,
men who created and carefully fostered their own dynamic,
nontheatrical uses of the filmic art. Robert Flaherty,
who was acclaimed by his friend and fellow-explorer Peter
Freuchen as "'The great name in Canadian subarctic exploration,'" completed his first film in 1922 at the age of 40.
He called it Nanook of the North. The film was about the
personable Eskimo hunter, Nanook of the Canadian Arctic.
Frances Flaherty notes:

> It was the first film of its kind, made without
> actors, without studio, story, or stars, just of
> everyday people doing everyday things, being them-
> selves. What is the secret of this very simple
> film? What is there about it that makes it en-
> dure? For commercially it is probably the most
> long-lived film that has ever been made. . . .
> When Nanook and Nyla and little Alegoo smile out
> at us from the screen, so simple, so genuine and
> true, we too, become simple, genuine, true. . . .

It can become that profound and personally lib-
erating experience we call "participation mystique."[22]

Returning momentarily to the Fogo experiment, we discover:

> The final result was 28 short films adding up to
> a total of 6 hours, each centered round a person-
> ality or an event rather than an issue; each ex-
> pressing an aspect of life on Fogo Island. . . .
> Through looking at each other and themselves,
> Fogo Islanders began to recognize the commonality
> of their problems as important, they began to
> become conscious of their identity as Fogo Island-
> ers; they discovered that preserving the Fogo en-
> vironment mattered to nearly all of them.[23]

At the urging of a friend that she select and "hammer
home" one word to describe her husband's work, Frances
Flaherty chose an explorer's word, "nonpreconception":
"Non-preconception is the pre-condition to discovery, because it is a state of mind. When you do not preconceive,
then you go about finding out. There is nothing else you
can do. You begin to explore." She then quotes her husband as saying, "'All art is a kind of exploring. To discover and reveal is the way every artist sets about his
business.'" She makes reference to L. L. Whyte's comment
in his book, The Next Development of Man: "Discovery is
the essence of social development, and a method of discovery
its only possible guarantee."[24] Certainly, on Fogo Island
40 years later there was nothing altogether new under the
sun: there was simply the rediscovery of the way to develop socially through the stimulus of filmic characters
interacting with their audience, creating a participation
mystique. A seed for filmic revolution was dropped when
Flaherty invited Nanook to participate in the decisions
of production. Frances Flaherty notes:

> Robert Flaherty is known as "The Father of Docu-
> mentary." And it is true that he was the first to
> fashion his films from real life and real people.
> But a Flaherty film must not be confused with the
> documentary movement that has spread all over the
> world, for the reason that the documentary move-
> ment (fathered not by Robert Flaherty but by a
> Scotsman, John Grierson) was from its beginning

all preconceived for social and educational purposes, for propaganda, and as Hollywood precon-
ceives, for the box office. These films are
timely, and they serve, often powerfully and with
distinction, the timely purposes for which they
were made. But there are other films, and the
Flaherty films are among these , that are timeless. They are timeless in the sense that they
do not argue, they celebrate. And what they
celebrate, freely and spontaneously, simply and
purely, is the thing itself for its own sake.[25]

If a PACT production unit can bypass argument to "celebrate, freely and spontaneously, simply and purely" the
local community for its own sake then quite possibly, with
the new technology, the Flaherty-Grierson influence will,
over time, loom large, indeed.
    In 1929, seven years after the arrival of Nanook of
the North, another film of unique genre was screened to
warm applause by the British Film Society. It was John
Grierson's Drifters. Forsythe Hardy considers Drifters
in its long-term effect:

    Drifters aroused immediate interest because of
    both its subject matter and its technique. In
    the studio-bound British cinema, whose most dar-
    ing expedition was into a West End theatre to
    photograph an Aldwych farce on a Co-optimists'
    show, a film which drew its drama at first-hand
    from real life was something revolutionary.
    Grierson's simple story of the North Sea herring
    catch brought what were then new and striking
    images to the screen: what has become familiar
    to-day through a thousand documentary films had
    then the impact of startling discovery.[26]

As a result of studying the Russian directors Pudovkin
and Eisenstein (he helped prepare the version of Potemkin
distributed in the United States), Grierson employed the
techniques of symphonic structure and dynamic editing in
his own film. In The Fortnightly Review of August 1939
he wrote:

    The documentary film movement was from the begin-
    ning an adventure in public observation. It
    might, in principle, have been a movement in doc-
    umentary writing, or documentary radio, or docu-

mentary painting. The basic force behind it was
social not aesthetic. It was a desire to make a
drama from the ordinary to set against the pre-
vailing drama of the extraordinary: a desire to
bring the citizen's eye in from the ends of the
earth to the story, his own story, of what was
happening under his nose. From this came our
insistence on the drama of the doorstep. We
were, I confess, sociologists, a little worried
about the way the world was going. . . . We were
interested in all instruments which would crystal-
lise sentiments in a muddled world and create a
will towards civic participation.[27]

It goes without saying that Grierson would be interested
in the combination of portapak, videotape, and cable. In
the Spring 1972 edition of the Newsletter Challenge for
Change Societe Nouvelle, at the end of an article by
Grierson there was a note from editor Dorothy Todd Henaut
which began:

John Grierson, founder of the National Film
Board, a man of incredible vision and commitment,
had a lively interest in the Challenge for Change
program; he stimulated, criticized and challenged
our ideas.
Shortly before this issue of the newsletter
went to press, Grierson died, at the age of 73.
We mourn the loss of our good friend.[28]

How far back in time does the idea reach, the idea
of holding a mirror up to life? The idea of self-education? Writing in the Canadian Affairs issue of June 15,
1944, the father of documentary film had this to say:

The main thing is to see this National Film Board
as a service to the Canadian public, as an attempt
to create a better understanding of Canada's present, and as an aid to the people in mobilising
their imagination and energy in the creation of
Canada's future. . . . A country is only as vital
as its processes of self-education are vital.[29]

As Forsythe Hardy points out, Grierson was hardly content with simply holding up a mirror:

Grierson, who combines the zeal of a practical
reformer with the imagination of a creative work-

er, is as well equipped as any man to hold the
aesthetic case; but to use his own words, he has
resolved his difference in the idea that a mirror
held up to nature is not so important in a dynamic
and fast-changing society as the hammer which
shapes it. . . . It is as a hammer not a mirror
that I have sought to use the medium that came
to my somewhat restive hand.[30]

About half the practitioners had been heard from in
a discussion at the Memorial University seminar about the
process of social change in relation to native Canadians
when Art Blue of the department of Indian studies, Saska-
toon campus, University of Saskatchewan, made his statement. In part he said:

You say to me, I don't want to be the leader. I
don't want to go into communities and change them.
I want to allow them to change for themselves.
Don't fool with me. Don't play with idle words
and talk meaningless nonsense. You have ideas
and you have meaning, and if you have meaning
you must also have this vision. You cannot go
empty or you would not go at all. Don't play
with me; don't play with the people. You must
also become a follower because once you are
there, you must follow the people, you must follow their meanings, you must help them to make
it paramount. But you must share with them your
vision, or otherwise you walk a lonely road without any end.[31]

And so the evidence is in. It is a hammer, not as a
mirror, that PACT can create social change, can stimulate
social animation. It can and should be used as a hammer.
It is, therefore, to some degree, a cybernetic process of
synthetic reproduction of men's intelligence, shared intelligence with the occasional objective of reaching a consensus.
"The point I want to emphasize," stated Lee Thayer
in his Mexico City speech (in which he made no mention of
cable television),

is that intercommunication is a process which can
occur only between or among people. Television
or radio may inform, but can in-form those they
thus reach only to the extent that such informa-

30

tion can be and is confirmed by significant human intercommunication. The "media" may incite
informing alternatives, but they cannot effect
those alternatives. Communicational realities
and the human institutions which are built out of
them are ultimately the products of intercommunication between people, not of the mass production
and mass distribution of messages.[32]

"Unquestionably, with one tenth of the population actively involved in making programs," asserts Sandra Gwyn,
"the Normandin experiment in community TV has been more
successful than any other."[33] The Normandin experiment
began in September 1970 and again a serendipitous factor
was a key motivator. Since a film crew had shot scenes
about forestry in the Normandin area, the local cable operator suggested that some of the unedited footage be cablecast on his system. In the words of National Film Board
producer Louis Portugais: "'People became very interested,
particularly the local school board, and they asked Societe-
Nouvelle for some help.'"[34] As a result, help came and
worked very intensively with a training program for a year
before turning over responsibility for continuation of the
idea to the local enthusiasts. The social animator and
sound engineer left with their six portapaks. From the
seeded community, comprising three villages, with a population of about 10,000, 10 percent or 1,000 became active
with PACT. Resources were developed within the community
to replace the seeding equipment. Managing those resource
people is a program committee, a 15-member board of directors. There are now three systems, each with a full-time
social animator, employed by the district school commission. According to Gwyn, "Each system produces between
one and two hours of community programming nightly. Rat-
ings for these have reached as high as 70%."[35]
      A report entitled Urban Cable Systems prepared by the
MITRE Corporation has this to say about community communications:

      All cities, including Washington, suffer from loss
      of community, disruption of life patterns, and so-
      cial alienation. This malaise is one causative
      factor in the more tangible problems of the city.
      Local and community dialogue, made possible by
      cable, offers one tool that might help to at least
      partially rebuild a sense of neighborhood, commu-
      nity, and identity. The MITRE system plan gives

31

high priority to this need by creating local
studios throughout the city, by recommending
strong financial support for programming, and by
providing many channels for public access, some
free and some on lease at moderate fees.[36]

"American democracy," in the opinion of Herbert E.
Alexander, "can be defined as government for the people
most effectively heard."[37]
Also in the MITRE report, which was released in May
1972, in a brief paragraph with the heading "Community
Programming" there is a recommendation for the employment
of social animators, but no mention of the source of this
idea. How the "enlargement of the idea" might manifest
itself at the urging of the MITRE Corporation is given
here:

One way transmission experiments would involve
substantial testing of community programming,
which is one of cable's greatest areas of potential benefit. A number of cable systems have
made channel space available, either without charge
or on a leased basis, to individuals and community
groups. Results, however, have sometimes been
minimal in terms of the quality of the material
and the public interest in the programming. The
problem is to provide both technical and production assistance to a whole stratum of the citi-
zenry that has previously had little or no direct
access to television, and knows very little about
how to use it [as at Normandin]. The demonstration project would therefore include the establishment of five Production Assistance Teams, who
would make themselves available to all community
cablecasting studios. These teams would help individuals and community groups to turn their
ideas and their interests into effective visual
programming.[38]

Other records in the literature concerning early Canadian experiments are of particular note. As early as
1968 the idea of a VTR production assistance team was
tested in the Challenge for Change project undertaken by
Dorothy Todd Henaut and Bonnie Klein in cooperation with
the Comite des Citoyens de Saint Jacques, "a dynamic citizen's organization in one of downtown Montreal's many poor
areas. . . . The main job of the Information Team to which

32

the VTR group was attached" according to Henaut and Klein, was "to sensitize the inhabitants of the area to their common problems and to communicate the Committee's hope that together they can act to change their situation."[39] As a result of their experience with the portable video taping unit, Henaut and Klein contributed considerably to the elaboration of the PACT idea. In their recommendations they commented:

> Hopefully, by using the 1/2" video equipment enough, a citizens' group could eventually propose to their local TV outlet that they make their own programs about themselves and their programs, to inform the population-at-large about their lives and aims, and to help bring about needed changes.[40]

They ended this statement by saying, "Unfortunately, 1/2" video cannot be transferred to the 2" broadcast video with any degree of technical satisfaction, for the moment. Perhaps technological advances will overcome this obstacle in the near future."[41]

"Hopefully" is a word worn ragged in recent times. Saturation of conversation with this expression seemed to begin about the time the "hopeless" wars on poverty, the North Vietnamese, and racism in the United States began. The poor people's experiment in Montreal with a portable VTR unit did, however, possess such qualities as might inspire more than a modest amount of hope. Now that the Henaut-Klein hope that a technological breakthrough would make it possible for the poor to gain access to a local TV outlet has been realized, a number of experiments in other communities across Canada and the United States have been undertaken and reported on. Beginning on November 9, 1970, Thunder Bay Community Programs went on the cable with four hours of programing each week. A crew of six graduates from a summer training project in community television launched this experiment that was to serve as a model that could be used to persuade the CRTC and others that such an activity would be worthy of official support. Henaut offered this comment on the programs:

> The Thunder Bay project has been put in a pigeon-hole labelled "failure." Yet without their initial reflections and drive--and their development of the Charter Board theory--community television would have died before it was born. The high de-

gree of technical expertise that was developed,
and the precedent of putting half-inch VTR on
cable, are important milestones in the exploration of new uses of media. The lessons learned
in Thunder Bay are important guides for future
development in the theory and practice of citizen
access to media. We have much to thank them for.[42]

One of Societe Nouvelle's latest contributions to in-
novation in community communication is the Montreal center
of videotape production, instruction, viewing, and distribution called Videographe.

Teams of "ordinary citizens" are learning how
to program up to 60 minutes of television, based
on their own program ideas. They are learning
to conceptualize, to film, the discipline of
editing, the uses of sound effects, music and
commentary as punctuation for their productions.
The facilities of Videographe are open,
through a program committee, to any group that
has formulated a clear idea of something it would
like to say through VTR and of the people to whom
it would like to speak. Authors of approved projects get a production budget and necessary technical assistance.[43]

Elizabeth Prinn, assistant editor of Newsletter Challenge for Change ociete Nouvelle, visited Videographe and
was reminded of European towns of centuries ago when the
artist and the people were much closer together, often
gathered under one roof: "Videographe is a contemporary
'public square.'"[44] To the town-meeting-town-pump analogue,
then, can be added that of the "public square." In the
years ahead when historians set down the first chapter of
PACT, most assuredly the original editor of the Challenge
for Change newsletter, Dorothy Todd Henaut, should be cited
as a foremost pioneer contributor to l'esprit d'aventure
of the PACT movement. In the final paragraph of an article entitled "Galloping Videoitis" Henaut attempted to answer the question, "But who is it that is expressing himself: the guy behind the camera or the guy in front of
it?" (This article, it is interesting to note, appeared
in the issue of the newsletter containing obituary statements honoring the Thunder Bay Programs.) She wrote:

I have this utopian dream, wherein as the pollu-
tion and smog slowly lift, and the fires in the

ghettos die down, fish jump in the streams once
more, greenery is renewed, people sing in the
streets, one catches glimpses everywhere of a
cable-VTR crew, composed of three people:
Johnny Appleseed, Cesar Chavez and a little old
lady in running shoes. Cheers![45]

Another experiment that has made a major contribution
to the enlargement of the PACT idea is the rather ambitious
program undertaken by the Metro Media Association of
Greater Vancouver. Getting under way in the summer of
1971 with a portapak journalistic service to metropolis
Vancouver, Metro Media has since developed at least four
unique contributions to the enlargement of the PACT idea:

[1.] Using our own resources and other resources
in the community we are working to build local
communication networks that are readily accessible to citizens. [2.] We have . . . also started
programming on broadcast television and have also,
through our support of Neighborhood Radio and the
alternate press, worked to gain public access to
other media. [3.] We have . . . recently added
new video equipment to our resources, including
a 1 inch studio working on incorporating super 8
film and 35mm photography. . . . [4.] A rather
exciting development is the inception of Pacific
Visions a commercial subsidiary of Metro Media.
. . . Pacific Visions will initially provide the
unique sorts of skills Metro Media has developed,
to the private sector and educational institutions.[46]

It is further explained that the Pacific Visions relationship is similar to that used by the Canadian National Institute of the Blind to operate their commercial cafeteria
service, Caterplan, and in no way jeopardizes Metro Media's
nonprofit status or its eligibility for government and
foundation grants.[47]

HISTORICAL DEVELOPMENT OF PUBLIC
ACCESS IN THE UNITED STATES

A system known as Cable TV, Incorporated, serving
Dale City, Virginia (approximately 25 miles south of Washington, D.C.), provided a channel for the first PACT ex-

periment in the United States. As N. E. Feldman pointed
out in his Rand report for the Ford Foundation on Dale
City Television, the Dale City local-origination channel
was of particular interest because it was made available
to the Junior Chamber of Commerce for community programing on a full-time basis from December 1968 to early 1970.
"It thus appears to be," wrote Feldman, "the first commu-
nity-operated closed-circuit television channel in the
United States." The Jaycees accepted financial responsibility for the channel on behalf of the community. No
advertising was carried even when an urgent need to broaden
the financial base developed as replacements became necessary for worn out and damaged equipment. DCTV's community
was comprised of 13,000 to 14,000 residents on the cable,
comparable to the community of 10,000 in the Lake Saint
John area of Quebec province that was exposed to the pioneer Normandin experiment with portapak half-inch video
recorder units. Two small conventional cameras and a
one-inch recorder were used by DCTV. The experiment appears to have gained very little national or international
publicity in light of the fact that it was the pioneer in
PACT. No social animators were involved, but there was an
advisory board with representatives from 13 community organizations that applied traditional broadcast management
controls in the utilization of personnel, equipment, and
cable time.[48]

Following the failure of DCTV in February 1970 it was
not until July 1971 that another attempt was made in the
United States to provide a community with free access to a
television channel. That experiment has been conducted on
four channels provided by two cable systems in Manhattan,
New York. Although beset with numerous difficulties, such
as those evolving from defective technology and unmanage-
able geographic disbursement of citizens with common ethnic
interests, the experiment continues at this writing with at
least moderate success. In terms of enlarging the idea of
PACT the New York experience is a pioneer effort in the
setting of a great metropolis (80,000 subscribers) in the
United States. At the same time Metro Media was getting
under way in Vancouver. The New York City government
franchises for cable starts in Manhattan, granted in 1971,
which required each of the two grantees to set aside two
channels for community use, has to be another major contribution to the idea. Foundation money earmarked for
PACT development appeared for the first time in the New
York City and Vancouver experiments. Writing in the June
13, 1972, edition of the New York *Times* television critic
John J. O'Connor made this observation:

In order to use public access, the public must
be educated in the tools of the medium. That,
aside from the monthly charge for a cable hookup
in the home, requires centers equipped with video
equipment and administrative staff. The three
main centers--the Reilly-Stern Global Village, New
York University's Alternate Media Center and
Theodora Sklover's Open Channel--operate primarily
on "seed money" from foundations, and there is no
guarantee that those funds will continue to be allocated indefinitely.[49]

O'Connor concluded by noting that "funds for an equipment center in Washington Heights were recently voted by
the local community board."[50] In an earlier column O'Connor reported:

With a $275,000 grant from the Markle Foundation,
the Media Center opened its doors in mid-1971.
It is one of several organizations--Open Channel,
Global Village, Raindance--actively concerned with
programming for public access channels. In New
York, the center has used half-inch video-tape to
record everything from a three-day local school
meeting to activities of the Lion's Club, con-
stantly attempting to pass along technical capabilities to community and nonprofessional users.[51]

Originally a Southerner, George Stoney, director of
the Alternate Media Center at New York University, apprenticed in community television with Canada's Challenge for
Change group. AMC's plan includes mailing finished tapes
to all parts of the country "along with information about
video equipment and production costs. The tapes . . .
serve as 'core' or 'module' programs which can be adapted
easily to similar situations in new locations."[52] Other
community television production groups in the New York
City area are Automation House Studio, Space Video Arts,
Videofreex, Filmmakers' Cooperative, and People's Video
Theater. PVT employs the mediation process based on the
earlier Montreal and Fogo Island experiments: tape one
side in a conflict, and show it to the other; then tape
the second's response, show it to the original group or
both groups together, and possibly tape that response or
exercise other variations on the mediation theme. "If
you want to use the word 'pioneers,' it applies to them,"
says Michael Shamberg. "They have made real break-thrus

[sic] in understanding and developing community media in work they've done with the Young Lords, Panthers, Chicanos, and Indians."[53] Then Shamberg makes this interesting statement: "They've developed techniques of video mediation which essentially cool people out by taping them and letting them relate to one another through the medium."[54] Portable Channel of Rochester, New York, working with an original grant of $15,000 from the New York State Council of the Arts has also provided tapes to the Manhattan public access channels. NYSCA has also provided assistance to Global Village, Raindance, PVT, and Filmmakers' Cooperative.[55]

A new twist on the PACT idea was contributed in Reading, Pennsylvania. Berks TV Company (a subsidiary of American Television and Communications, the nation's third largest CATV company in number of subscribers) provided free office space and a telephone plus $6,000 worth of portapak equipment in a cooperative effort with AMC to "seed" a PACT center in Reading. This experiment was launched with approximately 50 volunteers and a social animator in January 1972.[56] The fact that "Open Channel staff members devote time to speechmaking, writing articles, to appearing at legislative and executive hearings on a city, state, and federal level in an effort to acquaint the public with the potentials of public access television,"[57] certainly makes yet another unique contribution to the enlargement of the PACT idea.

Michael Shamberg, author of Guerrilla Television and a former Time magazine writer, provides this autobiographical sketch of the Raindance Corporation, which has been a builder of the PACT idea in no uncertain terms:

> In late 1969 we started a group called Rain-
> dance. The name has several functions. First,
> because it signifies no specific product or pro-
> cess, it allows us to do anything. Secondly,
> rain dance is a form of ecological valid antici-
> patory design. And finally, Raindance is a play
> on R & D, or research and development, from which
> the Rand Corporation takes its name.
>
> The original purpose and idea for Raindance
> (which came from Frank Gillette) was to explore
> the possibilities of portable videotape, which
> was then less than a year old, and generally to
> function as a sort of alternate culture think-
> tank concentrating on media.

While Frank and a few others have left, our current configuration includes around a dozen people with sound business help and a more stable collective. Ideally Raindance functions as a support center for everyone's individual or collective projects.[58]

The newsletter <u>Radical Software</u>, founded by Beryl Korot and Phyllis Gershung as an adjunct of the Raindance Corporation, is the only PACT newsletter in the United States at the present time that approaches the circulation and informational value of <u>Newsletter Challenge for Change Societe Nouvelle</u>. The latter is issued three or four times a year without charge by the National Film Board and circulates approximately 10,000 copies each issue in Canada and 2,000 in the United States. The responsibility of publishing <u>Radical Software</u> rotates from one PACT group to another and a subscription rate is charged.

In <u>Guerrilla Television</u> Shamberg begins "a personalized history" of early public access activities in New York City with a reference to the influence of Marshall McLuhan, the "Magus of Media" who some years ago quietly emerged from Edmonton, Alberta, to make his mark on society as a colorful prophet in the 1960s. Foremost, he warned that man must step away from his obsession with the content of telecommunications and include careful analysis of its structure in searching for the real meaning of this powerful new medium. Shamberg writes:

In 1968, when Sony introduced the first portable video camera, Paul Ryan was doing his term as a conscientious objector from the Army as a research assistant to Marshall McLuhan at Fordham University in New York.
Paul claims that he got into videotape to figure out if McLuhan was right, for if he were then Paul would be able to decode accurately a medium that McLuhan hadn't touched yet. Thus, he borrowed the new videotape equipment from Fordham for the summer.[59]

"The whole content of <u>Guerrilla Television</u>," explains Shamberg, "means enhancing flexibility. On a personal level it can be used as a system of strategies for (information) environmental survival. At the cultural level, it's a way to re-direct resources and energy."[60] McLuhan protege Paul Ryan is thus identified as possibly the first person in the United States to experiment with the flexibility of the portapak on a scholarly level.

Dorothy Todd Henaut and Bonnie Klein have described how in a November 1968 meeting in Montreal with the Comite des Citoyens de Saint-Jacques they introduced the idea of the mediation process using the new portapak technology combined with the Fogo film process. A Comite des Citoyens information team was formed at the meeting, composed of six residents of the neighborhood in addition to the two National Film Board social animators.[61] It is instructive, too, to note this item by Bonnie Klein entitled, "Training," in the March 1972 Portable Channel newsletter of Rochester, New York:

> To share our ideas and experiences about cable's possibilities, Portable Channel is developing training programs, both long and short term, for the people who will be needed to develop community programming and channels. Geographical, racial, and special interest groups, who are now both excited and confused by cable, would select candidates to be trained in all aspects of community television production, through field work on actual problems and projects. The training will include exposure to other media as well, so that its usefulness will not be totally dependent on the development of local cable stations. A training planning committee is at work now, and anyone interested should contact Bonnie Klein at Portable Channel. Remember: OPEN HOUSE on THURSDAY NIGHTS at 7:30, or call for other times.[62]

Since she was a Challenge for Change alumna, there is reason to believe that Klein and Portable Channel were influenced by Vancouver's Metro Media program, which was expanding from video tape to a multimedia approach by this time.

Other early PACT producers in the United States were the Media Access Center at Portola Institute and Ant Farm in California. Like the Videofreex in New York State, the Ant Farm goes on the road for a while each year. This procedure is an ideal cybernetic device, in Michael Shamberg's opinion:

> You go out to communities and do videotaping; they pay you to come. Ideally you plug-in to existing hardware and show people how to use what they already have.

40

In smaller communities especially, people
are hungry for novelty. With a media bus you
can entertain them while you're there, and then
turn them on enough for them to want to set up
their own media system after you've left.[63]

The Videofreex (nine of them in all) worked with several
vans and VW microbuses outfitted for a final editing of a
production, thus allowing them to leave a copy of the tape
in the community. There is no evidence of Canadian experimenters taking the nomadic approach, with the exception
of Vancouver's Werner Aellen. Following is excerpted material from his report entitled, "Video Voyageur," which
appeared in the Metro Media newsletter:

From January 14, 1972 to March 20, 1972, I
have been living out of a suitcase visiting groups
and individuals working with video, either as a
social tool or for artistic expressions.
Prior to my departure I had been in correspondence with some individuals, most of whose names
I obtained from Mike Goldberg's excellent "Video
Exchange Directory" (IMAGE BANK) which is very
widely circulated and used by many people as a
comprehensive source of contacts. (It is interesting to hear Europeans remark with admiration
that it is mainly Canadians who take the initiative in developing communication's channels of information exchange; Americans also do it sometimes,
but mostly for money.)
My journey took me to Washington State, Oregon, Calif., New York, Holland, France, Switzerland, Germany, Denmark, Sweden, and England.
I gave tape sessions or sometimes workshops
at colleges, universities, art schools, group
studios or private houses. I then made arrange-
ments to view other people's tapes and wherever
possible copied samples of their work off a monitor.[64]

Aellen further states that Europeans were quite surprised
to learn that Metro Media had received support from the
federal government.[65] Also among the nomadic experimenters
with PACT were Dean and Dudley Evenson, who joined the
Raindance Corporation after some pioneering video explorations with their video van called the Fobile Muck Truck.[66]

USE OF PUBLIC ACCESS
BY RELIGIOUS LEADERS

To round out a historical review of the PACT idea based on the Newman model,* it is perhaps fitting that the closing words be devoted to a major sociological force in most communities of the United States and Canada: the force of religion. Religious interests have been active in PACT in both countries; this news item from Winnipeg appeared in the Unitarian Universalist World as early as June 1, 1971:

> Discussions at the 10th annual meeting of the Canadian Unitarian Council focused on a budget which included funds to investigate community television possibilities in Canada and to prepare experimental videotape for tv use as well as for communication between Canadian U U societies.[67]

By February 1972 the Winnipeg church had produced one program and Toronto had begun cablecasting a regular schedule of programs in addition to undertaking long-range planning for the use of the new means of holding communion with the greater community. Calling the production company UniVision, the Toronto Unitarians defined their objectives as follows:

> To encourage a better understanding among various ethnic, philosophical and cultural groups and individuals through the use of the television media . . . to assist community groups and individuals through various media . . . to see themselves and others in new and different ways.[68]

The introductory issue of a "newsletter for churchmen, educators, and community leaders" appeared in February 1972. Called Cable Information and edited by the Reverend S. Franklin Mack, supervisor of Cable Information Service of the National Council of Churches, the newsletter is essentially a condensed clipping service of cable-television public-access development here and in Canada. In his former position as deputy director of the Office of Communica-

---

*See Chapter 3 for a discussion of Cardinal Newman's ideas.

tions of the United Church of Christ, Mack worked with the
director of that department, the Reverend Dr. Everett C.
Parker, in helping to establish the legal right of the public to participate in decision making that affects what
it sees, hears, and reads. The United Church has continued
a vigorous program of informing the public of its right to
have a voice in determining what kinds of programs it shall
receive and produce via cable television.[69]
     A news item dated June 1972 announced that a video-
tape public access workshop would begin in a storefront
donated to the project by the Christian Reformed Church in
Hoboken, New Jersey.[70] There is "no reason in the World"
why churches shouldn't use cable television, stated an official of Theta Cable, a cable television system that
serves the Greater Santa Monica Bay Area and Santa Monica
Mountains in Southern California. Possible uses suggested
were telecasting of church services and publicizing of fund
raising benefits.[71] But this is pure entropy.
     Research studies that have been done on PACT are lim-
ited to three: the study of the Challenge for Change Lake
Saint John experiment conducted by the Quebec government;
the study of local programing across Canada by CRTC; and
the more recent study of 15 localities across Canada by
Interchurch Broadcasting. The last study, set to be completed in September 1972 but running well behind that dead-
line, is a cooperative effort by the National Catholic
Communications Center and broadcasting departments of the
Anglican and United churches of Canada, with Lutheran and
Presbyterian affiliations. The study has four objectives:

          (1) To collect, from national and local sour-
     ces, comprehensive information related to present
     and potential development of community program-
     ming on cable television.
          (2) to disseminate among informants data
     which may help them deal more effectively with
     their own communications goals--especially as they
     relate to community programming.
          (3) to bring together informants who may be
     valuable information and communication resources
     to each other, especially in community program
     development.
          (4) to analyse comprehensively all of the
     data and their interrelations--and to compile and
     publish a comprehensive report. . . . The report
     will explain principal factors in present state
     of community programming on CATV, project possi-

ble ways in which institutions, groups and individuals can help develop the field.[72]

Now that the PACT idea has enlarged to its present considerable magnitude, certainly it must be time to subject the idea to empirical study as an ongoing support process. The joint church study adds yet another strand to the warp and woof of this youngish community catalyst for social change. With the idea enlarged to its current state and fixed momentarily in the empirical scrutiny of social scientists, what is there to be discovered beneath the surface of its enlargement? At this point in its development, does the idea appear to be sound? In Newman's terms is the healthy growth of the idea endangered by corruptions? As the original Christian idea is now almost totally obscured by a multitude of corruptions, so the PACT idea most assuredly will be plagued by a similar onslaught as men strive to bend it to selfish whim and fancy.

The New Samaritan Corporation of Waterbury, Connecticut, in a press release dated March 27, 1972, announced an experiment in cable television systems that will bear considerable study by scholars of varied interests. Technical consultant for the corporation John J. Karl stated that one segment of the system would be used for experimenting with many home and public services that cable system operators have often discussed but never tried. "The whole system will be a pilot to demonstrate how cable can serve the public interest and will bring experts to Waterbury from other communities around the U.S. to see what is happening every step of the way."[73] The New Samaritan Corporation was formed in 1970 by the Communications Confederation of the United Church of Christ under the leadership of the Reverend Arthur E. Higgins with the objective of improving "the conditions of people in the community." The corporation's activities cover a broad range of community needs, but the cable venture is by far the boldest in design. The unusual consortium of church leaders, businessmen, and local citizens ventures to wire the entire city of Waterbury (35,000 homes) within three years, according to communications director Everett Parker of the UCC. A loan obtained from Aetna Life and Casualty for approximately $3 million will serve as the corporation's bankroll. Special consideration concerning subscriptions is to be accorded people on welfare, or those unable to pay a regular rate, "primarily because emphasis will be put on job training and helping Waterbury unemployed."[74]

In tracing the character development of the PACT idea it is found that service to the poor and downtrodden of a community is held to be of ultimate importance. PACT is a cause that seems to be held with ardor and faith by its adherents. It could, in essence, be defined as a religious idea with potential for universal acceptance and use. Broadcasting magazine was prompted to report, "The Alternate Media Center is almost messianic in spreading its gospel of the advent of the common man in the television that's soon to be upon us."[75] Ann Arlen has commented, "With few exceptions, people involved with program production for Public Access receive little or no pay. They are a dedicated lot."[76] They are working, perhaps somewhat idealistically, against "the Second Coming of Television" becoming little more than "the Second Come-on."

As for the universality of the PACT idea, experiments in Carpentras, France, with the portable video-mediation process, in which each group of citizens--the young and the old--made a videotape presenting some arguments, after which all the people in the community were invited to assist in the debate, had this result: the antagonists were brought together in the end "a trouver un terrain d'entente."[77]

MILESTONES IN MAKING FILMS AND
TELEVISION PROGRAMS WITH THE PUBLIC

As mentioned earlier in this chapter and in the PACT chronology (see Appendix A), the idea of public access to the media, or the participation "mystique," was first conceived and executed by the genius of Robert Flaherty. It resulted that the penetration of the barrier to a subject's participation in the design of a film document by Flaherty's Canadian Eskimo friend Nanook was to be the revelation of an exceptionally viable idea that would somehow lie dormant for the next 42 years. There is this one exception mentioned by John Grierson:

> The basic tendency of the Challenge for Change program is to follow decently in the original cinema-verite tradition which the English documentary people associate with Housing Problems (c. 1936). With that film there was talk of "breaking the goldfish bowl" and of making films "not about people but with them."[78]

But finally, like Rumpelstiltskin, the idea did emerge again, in the work of the French Canadian National Film Board producers Fernand Dansereau and Raymond Garceau. Between 1964 and 1966 they turned out 24 films to be used by the Bureau of Agricultural Planners for Eastern Quebec Province in their work as socioeconomic animators around the town of Mont Joli. As a result of the spokesmen for the population expressing themselves in front of the camera, these films exposed an inventory of the socioeconomic problems of the region and undertook to help the local population develop a global awareness, including all of the constant factors present in the situation in which they found themselves.[79] This second leg of the public access evolution has not received any coverage in the Challenge for Change newsletters to date. Rather, this story appears in the Societe Nouvelle newsletter, Medium Media. According to Challenge for Change the "second leg" does not occur until the Fogo Island experiment conducted by Colin Low in cooperation with the Extension Service of Newfoundland's Memorial University in the summer of 1967. Whether the National Film Board's Fogo experiment represents the second or third leg of the development may not be clear, but it is crystal clear that Colin Low's rediscovery stands as a monumental milestone in the evolution of public participation in the processing of media. The medium employed by Low and his co-producers the Fogo Island people was film; first Nanook the Eskimo hunter, then French Canadian farmers, and finally English-speaking Canadian fishermen obtained public access through the medium of film. With the advent of the Fogo Island experiment in 1967 the viability of the public access concept accelerated dramatically. That was also the year of the Canadian Centennial Expo's dazzling film exhibition that included the National Film Board production Labyrinth, co-produced by Colin Low. The War on Poverty had been mobilized, suggesting a new awareness of human sociological needs that would lead to search for new technology for inexpensively publishing the communication of the masses rather than solely the formula fare of the media elite. By the following year Sony Corporation had introduced the portapak camera and playback videotape recorder unit that has now become a mainstay in the viability of public access.

In 1970 two rather ambitious attempts to develop public access to cable television evolved in Canada. These were the previously mentioned Town Talk venture in Thunder Bay, Ontario, and the Normandin experiment in the Lake St. John area of Quebec. The Town Talk venture lasted less than a year. Thunder Bay Community Programs producer Chad Hannah contributed this observation to the Challenge for Change newsletter:

> I suspect that our greatest failing was our inability to arouse interest and enthusiasm locally for the concept of community television. I've already mentioned that time was a factor (perhaps a rationalization). After experiencing Thunder Bay Community Programs and later acting as a consultant to Project Unicity in Winnipeg in the fall of last year [1971], I have many reservations about community cable. Yet I still think the idea can work.
>
> Perhaps for the moment community television is best described as an apology expressed by a nation unwilling to allow public access to the established and powerful broadcast media.[80]

"Unquestionably, with one tenth of the population actively involved in making programs," wrote Sandra Gwyn in her report on the 1972 film, videotape and social change seminar in St. Johns, Newfoundland, "the Normandin experiment in community TV has been more successful than any other."[81] According to Gwyn's analysis there appear to be three basic reasons for this success:

--The area had a considerable degree of community consciousness. "This exists in rural Quebec," [the National Film Board's Societe Nouvelle social animator Louis] Portugais suggested, "because half of our population is in Montreal, and others feel somehow like 'employees' of Montreal."

--The local cable operator was receptive and enthusiastic.

--Local educational authorities have been extremely co-operative. Sensitized to television's potential by the TEVEC experiment in 1968-69 (an adult educational television blitz in the Lac St. Jean region), the District School Commission now assumed much of the responsibility for the project. Much of the equipment used by the community is the property of the board and is made available to the community virtually twenty-four hours a day.[82]

"Significantly," noted Gwyn, "Portugais came down strongly in favor of 'group viewing' of community television rather than viewing at home.

"'The Normandin cable system services 5 villages, of which three are hooked into the system. Recently they asked the CRTC for permission to install a transmitter to extend the system. I must confess that I was against it, said Portugais. 'Community television is a process--not a product--and it works better with a group.'"[83]

Elsewhere in Canada, according to Gwyn's determination, "the development of community programming has been mostly hit and miss."[84] Actually there are now quite a number of experiments to observe. A survey published in the summer of 1972 by CRTC revealed that 54 systems were doing at least two hours of community programing each week; 40 produced between two and five hours, and 25 systems telecast more than five hours of programing per week.[85] Many of the smaller urban areas must be quite alive with public access hits in order to maintain a production output of the magnitude detailed above.

New York City's Richard Galkin, president of Sterling Manhattan Cable Company (a $20-million loser on the operation last year) now views PACT with extremely guarded optimism:

> Cable could have been a significant force for social change. . . . Eight years ago, when I was a young guy I saw myself as participating in a really meaningful revolution. It may be hard to picture me as a revolutionary, but I saw revolutionary things, town meetings, real participatory democracy. . . . I used to dream about the day we'd have town meetings on cable where all the constituents could vote, and even national referendums electronically compiled on significant issues like should we bomb Vietnam?[86]

Galkin's concept of cable as a force for social change seems to be thoroughly interwoven with pushbutton "blue sky" services, which tend to have more appeal to a business mentality.

By the end of 1971 the two Manhattan public access channels were using an average of five to six hours each day, according to a March 1972 report for the Center for the Analysis of Public Issues, which also noted, "The problem of financing is key. The different programming experiences of TelePrompTer and Sterling demonstrate that free

48

channel time, by itself, is not enough to generate significant spontaneous usage." Hitting hard on the need for adequate and continuous financing the report further stated, "it seems clear that a no-subsidy policy now will stifle public access usage before it has a chance to develop its own broad base of support."[87]

In response to the need for a firm financial base for public access production comes this proposal of considerable historical portent, reported in the May 28, 1973, issue of Broadcasting:

> Support within the cable industry is mounting for a proposal now before the FCC that would permit the establishment of nonprofit community organizations to administer and control financing of public-access channels provided by CATV systems. Comments in support of theproposal from TelePrompTer Corp., the nation's largest cable operator, and the National Cable Television Association are now on file at the commission.
>
> The TelePrompTer and NCTA briefs are directed to a petition for a declaratory ruling filed several months ago by Open Channel, a New York-based nonprofit group that produces public-access programming and aids other community groups around the country in obtaining similar facilities.
>
> NCTA told the commission that it supports "the general thrust" of the Open Channel petition, with two stipulations: that financing for the proposed community-controlled administrative bodies be financed by no more than 5% of a respective cable system's gross revenues (this sum being allocated to the community group from the system's annual franchise fee, which the commission has ordered not to exceed 5% of the system's gross revenues), and that the proposal be implemented on an experimental basis until 1977, and only in the top-50 markets.[88]

1973 was a year of "conscience-raising" and "hands-on" cable television workshops. Workshops with a public-access orientation sprang up in Lincoln, Nebraska; Atlanta, Georgia; Aspen, Colorado; Claremont, California; Cleveland, Ohio; Indianapolis, Indiana; Kansas City, Missouri; and at Yale University. On April 29, 1973, the Washington Community Video Center opened a new storefront community video workshop and viewing center. The center was established for the purpose of initiating workshops in video production, cable television, related subjects, and to explore ways of using media for community development and information.[89] Washington does not have cable television, but when cable arrives there, it should become the final extension of an already viable video community schooled in the ways of public access. And so it might be with most of the other workshop hosts listed above. In these workshops ordinary citizens of the United States and Canada should have an opportunity to learn about a new instrument that can be used to "crystalise sentiments in a muddled world and create a will towards a civic participation." A major catalyst in creating that will has been the Alternate Media Center at New York University. On May 14, 1973, Broadcasting announced that the John and Mary R. Markle Foundation, the original benefactors of the center, had given an additional grant of $325,000. To date AMC has helped set up cable television public-access production outlets in New York; Cape May, New Jersey; Reading, Pennsylvania; Columbus, Indiana; Bakersfield, California; Orlando, Florida; DeKalb, Illinois; and Wildwood, New Jersey.[90]

On July 4, 1973 the Edmonton Journal quoted Canadian Communications Minister Gerard Pelletier in his tabling of CRTC's annual report in the House of Commons as saying that Canada will remain, "a world leader in the number of broadcast services per capita." The Journal article went on to state: "At the end of the 1972-73 fiscal year, March 31, there was a licensed broadcast undertaking for every 13,500 residents, compared with one for every 18,000 in the U.S." And finally, "The CRTC encourages decentralized and locally controlled broadcasting as one of the principal foundations of the broadcasting system."[91] There is, then, considerable evidence supporting the belief that the Canadian government understands and applies the concept of negentropy to the medium of television.

NOTES

1. Forest H. Belt, "Cable TV--Where It Is and Where It's Going," Popular Electronics (January 1971), p. 26.
2. Ibid.
3. Ibid., p. 28.
4. Arthur C. Clarke, Voices from the Sky: Previews of the Coming Space Age (New York: Harper & Row, 1965), p. 139.

5. Charles Tate, ed., Cablevision in the Cities (Washington, D.C.: The Urban Institute, 1971), pp. 16-17.

6. Hubert J. Schlafly, The Real World of Technological Evolution in Broadband Communication, a report prepared for the Sloan Commission on Communications (September 1970), p. 37.

7. Ibid., p. 37.

8. Mary Alice Mayer, "An Historical Study of the Issues and Policies Related to the Educational Application and Utilization of Community Antenna Television," Ph.D. dissertation, Northwestern University, Evanston, Ill., 1969.

9. E. Stratford Smith, "The Emergence of CATV: A Look at the Evolution of a Revolution," Proceedings of the Institute of Electrical and Electronic Engineers 58 (July 1970): 968.

10. League of Oregon Cities, "Cable Distribution of Television: A Survey of Practices in 104 Oregon Cities" in Policy and Practice in Oregon Cities (June 1967), p. 1.

11. CRTC, Canadian Television in Canada (The Commission, January 1971), pp. 5, 21.

12. "A Short Course in Cable, 1972," Broadcasting, May 15, 1972, p. 45.

13. Fernand Dansereau, Newsletter Challenge for Change 1 (Winter 1968-69).

14. Ibid., pp. 4-5.

15. Ibid., p. 5.

16. David Gee, Newsletter Challenge for Change 1 (Winter 1968-69): 7.

17. Jagsit Singh, Great Ideas in Information Theory, Language and Cybernetics (New York: Dover, 1966), p. 1.

18. Sandra Gwyn, "A Report on a Seminar Organized by the Extension Service, Memorial University of Newfoundland, St. John's, Newfoundland, March 13-24, 1972," Film Video-Tape and Social Change, pp. 4-5.

19. Ibid., p. 5.

20. Ibid.

21. Ibid., p. 2.

22. Frances Hubbard Flaherty, The Odyssey of a Film-Maker: Robert Flaherty's Story (Urbana, Ill.: Beta Phi Mu, 1960), p. 12.

23. Gwyn, op. cit., p. 5.

24. Flaherty, op. cit., p. 10.

25. Ibid., p. 11.

26. Forsythe Hardy, ed., Grierson on Documentary (London: Collins, 1946), p. 14.

27. John Grierson, "Documentary Film in Canada," The Fortnightly Review 146 (August 1939): 121-22.

28. Dorothy Todd Henaut, note on John Grierson, "Memo to Michelle about Decentralizing the Means of Production," Newsletter Challenge for Change Societe Nouvelle 1 (Spring 1972): 5.

29. Hardy, op. cit., p. 24.

30. Ibid.

31. Gwyn, op. cit., p. 29.

32. Thayer, "On Human Communication and Social Development," Economies et societes: La communication II, Vol. V (Geneva, Switzerland: librairie Droz, September 1971), p. 51.

33. Gwyn, op. cit., p. 25.

34. Ibid., p. 24.

35. Ibid., pp. 24-25.

36. MITRE Corporation, Urban Cable Systems (Washington, D.C., 1972), p. II-5.

37. Herbert E. Alexander, "Communications and Politics: The Media and the Message," Law and Contemporary Problems 34 (Spring 1969): 255.

38. MITRE Corporation, op. cit., p. II-53.

39. Dorothy Todd Henaut and Bonnie Klein, "In the Hands of Citizens: A Video Report," Newsletter Challenge for Change 1 (Spring-Summer 1969): 2-3.

40. Ibid., p. 5.

41. Ibid.

42. Dorothy Todd Henaut, "Town Talk: Community Intercommunication in Port William and Port Arthur," Newsletter Challenge for Change Societe Nouvelle 1, no. 8 (Spring 1972): 8, 10.

43. Elizabeth Prinn, "Vive le videographe!" Newsletter Challenge for Change Societe Nouvelle 1 (Spring 1972): 18.

44. Ibid.

45. Dorothy Todd Henaut, "Galloping Videoitis," Newsletter Challenge for Change Societe Nouvelle 1 (Spring 1972): 3.

46. Chris Pinney, "Chris Pinney Looks at Our First Year in Retrospect," Metro Media Print-Out 1 (July 1972): 7-8.

47. Ibid., p. 8.

48. N. E. Feldman, Cable Television: Opportunities and Problems in Local Program Origination, a Rand report for the Ford Foundation (September 1970), pp. 10, 12.

49. John J. O'Connor, "TV: Added Exposure for Public Access on Cable," New York Times, June 13, 1972.

50. Ibid.

51. John J. O'Connor, "Public Access Experiments on Cable TV Advancing," New York Times, June 6, 1972.

52. Center for Analysis of Public Issues, New York CATV Supplement, supplement no. 2 (August 1971), p. 1.

53. Michael Shamberg and Raindance Corporation, Guerrilla Television (New York: Holt, Rinehart and Winston, 1971), p. 17.

54. Ibid., p. 19.

55. Center for the Analysis of Public Issues, Public Access Channels: The New York Experience, a report for the City of New York (March 1972), p. 4.

56. Broadcasting, May 1, 1973, p. 46.

57. Shelley Fieland, "Questions and Answers About Open Channel," Feedback Feedforward (July 12, 1972).

58. Shamberg, op. cit., p. 37.

59. Ibid., p. 10.

60. Ibid., p. 91.

61. Henaut and Klein, op. cit., p. 2.

62. Bonnie Klein, "Training," Feedback Feedforward, March 1972, p. 4.

63. Shamberg, op. cit., p. 92.

64. Werner Aellen, "Video Voyageur," Metro Media Print-Out 1 (July 1972): 5.

65. Ibid.

66. Shamberg, op. cit., p. 12.

67. Unitarian Universalist World 2 (June 1, 1971): 1.

68. Ibid. 3 (February 15, 1972): 1.

69. "Cable Television Lures Churches," Evening Outlook, June 17, 1972, p. 10.

70. "Hoboken Will Open Videotape Workshop," The Jersey Journal and Jersey Observer, June 1972, p. 10C.

71. "Cable Television Lures Churches," p. 10.

72. Cable Information 1 (July 1972): 4.

73. Jack Gould, "Waterbury Officials Planning for First CATV City," The New York Times, January 20, 1972, p. 86.

74. Ibid.

75. "Open Access: What Happens," Broadcasting (May 1, 1972), p. 46.

76. Ann Arlen, "Will Public Access be the Second Coming of Television," Foundation News, May/June 1972, p. 28.

77. Marie-Therese Guichard, "Le cable cassera le monople," Le Point 11 (December 4, 1972): 35.

78. John Grierson, "Memo to Michelle . . . ," p. 4.

79. Medium Media, no. 1 (Autumn 1971), p. 4.

80. Chad Hannah, "Thunder Bay Community Programs," Newsletter Challenge for Change Societe Nouvelle 1, no. 8 (Spring 1972): 10.

81. Gwyn, op. cit., p. 25.

82. Ibid.

83. Ibid.

84. Ibid.

85. Broadcast Programmes Branch, Canadian Radio-Television Commission, "The State of Local Programming" (Summer 1972), p. 11.

86. Robin Reisig, "Cable TV: The Light That Failed?" Village Voice, March 23, 1973, p. 90.

87. Center for the Analysis of Public Issues, Public Access Channels: The New York Experience, a report for the Fund for the City of New York (New York: The Center, March 1972), p. 12.

88. Broadcasting, May 28, 1973, pp. 51-52.

89. Cable Information 2, no. 4 (June 1973): 2.

90. Broadcasting, May 14, 1973, p. 51.

91. "CRTC Pushes Local Talent," The Edmonton Journal, July 4, 1973, p. 68.

3

FACTORS CORRUPTIVE
TO ENLARGEMENT OF
THE PACT IDEA

In striking a model that will aptly display some major elements of an analysis of PACT, it is first necessary to ask what PACT is. Something beyond the functional definition is needed here. It is obviously an idea but not an ordinary one; rather, it is a plan for action. There is evidence to suggest that one day the idea will assume the role of a governmental doctrine on an international scale as it spreads to meet with universal need and acceptance.

Charles Frederick Harrold stated that, "like Darwin, [John Henry Cardinal] Newman was less the inventor of a theory than its promulgator; both men took the general nineteenth-century notion of 'growth' and 'development,' and illuminatingly applied it in a special field of inquiry. That is not to say that there is any similarity between Newman's developmentalism and Darwin's evolutionism."[1] Harrold also noted, "Newman came to prefer the term 'enlargement,' which preserves the idea of original purity along with subsequent growths."[2] The surveys of PACT undertaken in this study are subjected finally to an analysis of the enlargement of the original idea in terms of the extent and purity of that enlargement. It was Newman's reasoning that:

A power of development is a proof of life, not only in its essay, but especially in its success; for a mere formula either does not expand or is shattered in expanding. A living idea becomes many, yet remains one.

The attempt at development shows the presence of a principle, and its success the presence of an idea. Principles stimulate thought, and an idea concentrates it.[3]

The analysis of the surveys begins, then, with identification of those principles that seem to have stimulated the original thought that produced the PACT idea. The surveys allow us to trade roughly the adventures of the idea in a variety of communities as it moves out helter skelter across North America. What do the surveys demonstrate the reception and utilization of the idea to be in the various communities in which it is encountered? Is there evidence in any instance of the idea being subjected to obvious corruptive factors? What appear to be some causes of any isolated cases of the idea failing to take root and develop?

On the process of development in ideas, Newman had this to say:

IT IS THE CHARACTERISTICS OF OUR MINDS to be ever engaged in passing judgement on the things which come before us. No sooner do we apprehend than we judge: we allow nothing to stand by itself: we compare, contrast, abstract, generalize, connect, adjust, classify, and we view all our knowledge in the associations with which these processes have invested it.[4]

Do the surveys reveal whence the idea gets its "motor" for moving from community to community, like a Johnny Appleseed with energy and direction for trekking ever onward across borders into strange new cultures uninitiated in portapak communion? Is there a "participation mystique?" In Harrold's analysis:

Newman suggests that ideas and doctrines have a way of moving about as if charged with a power of their own. They enter men's minds, and use them as instruments, even while men believe that they, as thinking beings, are producing and controlling the ideas. A germinal idea may rise in the mind of a simple fisherman of Galilee, it may find utterance in a few simple words. Then, as if it has a life of its own, it will expand and develop, it will progress in the minds of many men with consistency, resourcefulness, and creative power.[5]

And in Newman's own words:

> It will be interrogated and criticized by ene-
> mies and defended by well-wishers. The multitude
> of opinions formed concerning it in these respects
> and many others will be collected, compared,
> sorted, sifted, selected, rejected, gradually at-
> tached to it, separated from it, in the minds of
> individuals and of the community. It will, in
> proportion to its native vigour and subtlety, in-
> troduce itself into the framework and details of
> social life, changing public opinion, and strength-
> ening or undermining the foundations of established
> order. . . . And this body of thought, thus labor-
> iously gained, will after all be little more than
> the proper representative of one idea, being in
> substance what that idea meant from the first,
> its complete image as seen in a combination of
> diversified aspects, with the suggestions and cor-
> rections of many minds, and the illustration of
> many experiences.[6]

A PACT consultant, for example, looking back through
the pattern of development of the PACT idea, would be particularly curious as to the key factors that have contributed to the growth of the idea, and, equally important,
those that have negated the idea. The annotated bibliography and survey of the literature at the end of the present work should provide a substantial insight into those
factors that appear to contribute to the enlargement of
the PACT idea. But at this point there should be some
value in looking more closely at the major instances where
the idea seems to have gone astray.

The survey reveals that the idea, the concept of participation mystique, was the creation of Robert Flaherty
in 1922. Following Flaherty's _Louisiana Story_ in 1946 the
participation approach lay fallow for 20 years. In February 1967 Fernand Dansereau began production of a participative or social animative film with satellite films in a
community study of Saint Jerome, Quebec. At that time
Dansereau had this to say about film:

> The starting point for self-realization is the
> power to express oneself. Film gives this power,
> better than any other medium--it is the most di-
> rect medium of expression. It would be ideal if
> people could make this type of film themselves.[7]

Dansereau's humanistic approach was prompted by the very unhappy experience of National Film Board producer Tanya Ballantyne, who inadvertently caused extreme mental anguish in a family of subjects in her film on poverty, by using a traditional nonparticipative documentary approach. As a result of this incident, no Challenge for Change film has since been distributed without the subjects participating in the final editing process. Through her innocent corruption of the filmic process as applied to social animation Ballantyne triggered the "second coming" of participation mystique in the motion picture of man.

That this rediscovery has not yet attained universal acceptance is evidenced in the approach taken by Craig Gilbert in his 12-hour series produced for PBS, An American Family, which is about an "average" American family in Santa Barbara, California. The 1973 upper-middle-class Loud family of Santa Barbara has a great deal in common with the 1966 poverty-stricken Bakey family of Montreal, insofar as the members' experience with documentary film is concerned. This account of the Loud's experience with the nonparticipative mystique is taken from the Kansas City Star column by television critic Joyce Wagner:

> "We have absolutely been through hell with the critics," Pat Loud explained. "We've had such lurid publicity that I feel I'm justifiably upset about it."
>
> "When we went into it we thought it would be one big, great adventure," Bill Loud injected.
>
> "Yes," Pat conceded, "we were under the aura of educational television. We thought, 'Oh, boy, they're going to handle this right.' If it had been a commercial channel," she added, "we might have had second thoughts about it. We might have been more concerned about what we were getting into."
>
> "I think the thing that upset us the most," Bill Loud interrupted, "was that Craig elected to show the crises rather than the good times we had during the filming.
>
> "It seemed that if he had one good shot and 20 negative he would choose one of the negative."
>
> At this point, Gilbert interrupted to explain that the editing had "not been a one-man job. It was a joint decision by seven or eight people."

"Then why do people say we look like freaks
and monsters and zombies?" Pat Loud replied, not
without a tinge of hostility.

Gilbert launched into a lengthy dissertation
about the state of the nation and its citizens'
desire to alleviate unpleasantness from their
lives. But he never really answered the question.[8]

The local origination channel made available for community programing on a full-time basis to the Junior Chamber of Commerce of Dale City, Virginia, in December 1969
functioned for approximately one year. N. E. Feldman of
Rand offered two major reasons for the demise of the experiment: lack of financing and low-quality equipment.
By the end of the first year of operation the low-quality
equipment had worn out, and it was beyond the resources of
the community to provide substantial and continuous financial support. The one-inch videotape recorder caused considerable trouble, and working without a permanent studio
proved to be a serious handicap.[9] The Dale City experiment
preceded the introduction of the relatively inexpensive
half-inch Sony portapak system.

The first failure of a PACT experiment involving the
use of portapak occurred in Thunder Bay, Ontario. Although
there was a lack of diversified financial support for the
popular Town Talk public-affairs program series, funding
for the experiment in general was probably excessive to
the point that it obviated development of local resources.
Special projects director Dorothy Todd Henaut of Challenge
for Change had this to say about the experiment:

The crew were chosen only for their skills in
film and VTR, and they lacked training in the
techniques of community development. They had
neither the training nor the time to teach people
the use of VTR as a social tool. They were not
asked to make each technical training workshop
with local citizens serve as a building-block for
a strong media organization. Nor did they have
the time for building grassroots support.[10]

Certainly the elitist, nonparticipative production approach
developed by the six full-time staff members failed in the
essential function of building a productive liaison between
the community production team and the cable management.
Chad Hannah relates:

Resigned as we were to our untimely fate, we
agreed to meet with the regional operations manager for the cable company. The decisions were firm
but the chance to have a chat was too inviting to
pass up. It was clear from the start that the
famed meeting would go down as the "OK Corral" in
cabledom history.

Fortunately, we were in top form that after-
noon, our minds stropped to razor-like sharpness.
Our parries and jabs went unanswered, our foot-
work dazzled them, their arguments were dissected
with swift rapier-like thrusts, and our witticisms
were exquisite.

We had won the battle, yes, but we had lost
the war! The encounter had served only to refur-
bish our bruised egos.[11]

Both in Vancouver and in New York City the PACT idea
was naturally threatened with production elitism because
of the potential for monolithic growth. In both instances,
however, the need for decentralized communities is currently
being acted upon. The two franchises granted by Manhattan
require the addition of 20 headends or sources of origination, so it will become relatively simple to filter programs
to specific neighborhoods.[12] The Center for the Analysis
of Public Issues made this evaluation in March 1972 of a
city's role in safeguarding against corruptive factors
that are damaging to the enlargement of the PACT idea:

The regulatory problem facing the local franchising authority is an exceedingly subtle one, if
carried out properly. Conventional regulatory
problems--the quality of the signal, the subscription price, etc.--might easily be left to market
forces in the case of a fledgling industry such
as cable television. The city does need to pay
attention, however, to the kind of equipment being installed to ensure that it will be compati-
ble with the technological advances most likely
to occur in at least the relatively near future,
and should also take an aggressive and positive
role in developing experiments in public-access
and municipal usage. It must also become much
more involved than it has to date in problems of
sub-districting for neighborhood-access productions. And in problems of compatibility of
equipment and scheduling inconsistencies between

the companies which on several occasions have presented frustrating problems to would-be users.[13]

Among the factors that have been instrumental in cor-
rupting the healthy enlargement of the PACT idea, then,
are these seven:

1. Instability of programming as a result of relying
too heavily on one source of funding;
2. Failure of "seeding" personnel to develop a strong
liaison with the local cable ownership and management;
3. Failure of "seeding" personnel to develop grass-
roots support through production workshops;
4. The tendency in large metropolitan areas for the
development of monolithic, elitist production units comparable to those of the broadcast networks;
5. Insufficient available resources or potential for
the development of local resources to insure purchase and
maintenance of quality equipment;
6. Failure to provide a studio for PACT productions
only; and
7. Incompatability of equipment and systems in multi-
system markets.

NOTES

1. John Henry Newman, An Essay on the Development of
Christian Doctrine, Charles Frederick Harrold, ed. (New
York: Longmans, Green, 1949), p. xxviii.
2. Ibid., p. xxix.
3. Ibid., p. 173.
4. Ibid., p. 31.
5. Ibid., p. xix.
6. Ibid., p. 35.
7. Fernand Dansereau, Newsletter Challenge for Change
1 (Fall 1968): 10.
8. Joyce Wagner, Kansas City Star, February 25, 1973,
p. 4F.
9. N. E. Feldman, Cable Television: Opportunities
and Problems in Local Program Origination, a report by the
Rand Corporation for the Ford Foundation, September 1970,
pp. 10-12.
10. Dorothy Todd Henaut, "Thunder Bay Community Programs," Newsletter Challenge for Change Societe Nouvelle 1
(Spring 1972): 8-9.

11. Chad Hannah, "Thunder Bay Community Programs,"
Newsletter Challenge for Change Societe Nouvelle 1 (Spring
1972): 10.

12. George Gent, "Public Access TV Here Undergoing
Growing Pains," New York Times, October 26, 1971, p. 83.

13. Center for the Analysis of Public Issues, Public
Access Channels: The New York Experience, a report for
the Fund for the City of New York (March 1972), pp. 34-35.

4

**THE FUTURE OF
PUBLIC ACCESS
CABLE TELEVISION**

THE EFFECTS OF PACT

Two general criteria that serve as a basis for analyzing the data received from the PACT questionnaire (see Appendix B) respondents are:

1. What pertinent information, as contained in the responses to the questionnaire, have the respondents individually and collectively provided concerning the viability of PACT?

2. In what manner and to what extent, as elicited by the questionnaire, are city governments involved with planning for PACT?

A portion of the data is illustrated by a map of the 150 cities surveyed by the questionnaire, shown facing Chapter 1. The map provides a comprehensive view of cities responding to the questionnaire; responding cities having PACT; cities that planned to install cable systems in 1973; and cities currently having cable systems but no PACT. The essential consideration in the integration of the responses into an analysis is negentropy: in other words, is there in the response any evidence of the respondent city's contributing to the organization of PACT, or, to what degree does each response reveal an interchange of "atoms and molecules" by city governments singly and collectively with the input of the PACT idea?

Because of its importance to the design of this study, this additional insight into the entropy-negentropy concept by John Young should be helpful in illuminating the structure of the analytical model:

Studies using radioactive isotopes have
shown that even the bones of the body are not
composed of the same atoms permanently, but only
retain particular atoms for a limited time. Thus
although a living body maintains itself distinct
from its surroundings, it does this by taking
atoms and molecules from the surroundings and organizing them into a living shape. Mathematically,
this characteristic of life can be called negative entropy. The entropy of any system increases
as it becomes more disorganized, decreases as it
becomes more organized. A deck of cards with all
the suits and numbers in order has low entropy.
Shuffle the cards and the entropy increases.
Throw the deck in the air so that the cards spread
randomly all around the room and the entropy increases still more. Collect all the cards together and sort them into suits and correct number order and the entropy is decreased to its
original value. Living organisms all have the
property of taking up randomly distributed matter
and organizing it into a logical, living shape.
When this process of organization ceases, the organism can no longer be said to be alive.

Thus the primary aim of life appears to be
to maintain its state of organization and to interchange atoms and molecules with its surroundings
in the process.[1]

All the questions in the survey can generate information in terms of how extensively the PACT idea has been organized locally into "a living shape." Nine of the questions in the questionnaire have a built-in scale. The remainder tend to be either-or questions: either the city
government has interchanged some PACT atoms and molecules
or it has not. Many variables will be at work distinguish-
ing categories such as cities without cable; cities in the
top-50 U.S. television markets; cities in the 51 to 100
largest U.S. television markets; and Canada's 50 largest
cities. The value of reporting this study can be judged,
for example, by its apparent usefulness to a hypothetical
"seeding" agency that must determine priorities in terms
of the cities that might best be served by an enlargement
of the PACT idea in North America. The "seeding" agency--
for example, Alternate Media Center--might give top priority to a survey respondent revealed to have a cable operator friendly to PACT who has just been franchised. Other

data in that survey, however, might show that this particular respondent might very well have a cable operator who is friendly to PACT, but, at the same time might have no civic leaders sufficiently imbued with enthusiasm for the PACT concept to martial city resources around a public access experiment. The total number of respondents was 106 out of a possible 150 (a 71 percent return). Not all of these 106 responded with a questionnaire. Many of those mayors' offices in cities that did not as yet have cable chose to respond with a letter.

A half century ago Henry Ford wrote, "Every moral or social ideal indicates the pressure of better conditions which are trying to break through and become the rule of life."[2] PACT is considered by a growing number of people to be an excellent tool for facilitating the breakthrough in a community of better social conditions. This study has been directed, in part, to an examination of the rationale of PACT as a means of social animation or community therapy. There has been an attempt, then, in this study to delineate the sociological theory and philosophy of the public access idea insofar as it has potential for animating social cultural change. The major source of this theory and philosophy has been the National Film Board's newsletter, published by its social animation subsidiary, Challenge for Change. The newsletter has been published three or four times a year since the spring of 1968 and is circulated free of charge. In searching out the results of experiments with PACT the Challenge for Change newsletter has been most helpful. With the introductory issue of the newsletter, Cable Information, in February 1972 an excellent source of tidbit information about public access developments became available through a $10 per year subscription with the Cable Information Service of the Broadcasting and Film Commission of the National Council of Churches. In the introductory issue it was announced the monthly newsletter would carry "capsulized treatments of CATV developments; reviews and analyses of new books, commission reports, studies; updated listings of organizations and publications dedicated to informing and assisting community leaders; also innovative uses of cable, creative program origination, citizen action, feedback."[3] An important source of the rationale for PACT has been Michael Shamberg's Guerrilla Television (cited in various instances in this study). Another method used to obtain general information for the study was the inclusion of an invitation in the survey questionnaire, "Please enclose any additional information that you feel would contribute to this study

of public access (community) cable television in the United
States and Canada."

Discovery of the various public access experiments
referenced in this study was accomplished primarily through
exploratory correspondence with more than 100 individuals
and groups in the United States and Canada that are in some
way involved with the public access idea. These range
through cable operators, governmental agencies, private
foundations, cable information centers, and private citizens. In light of the mandate from the FCC to the major
cities of the United States to provide public access to
the cable systems they franchise, and of the strong urging
from the CRTC that Canadian cable systems provide public
access, the prime purpose of this study has involved surveying the offices of the mayors in 150 of the major cities
in North America for a measure of the cities' response.
The 30-question survey instrument, therefore, was designed
to ask in 30 different ways what, if anything, city hall
is contributing toward development of the PACT idea. Information garnered about PACT has been displayed throughout the text, in the development of each chapter, by means
of a map, a chronology, and an annotated bibliography. As
the prime purpose of this study is to update, expand, and
deepen the story of PACT in the United States and Canada,
one method employed to help maximize the response to the
questionnaire was to offer respondents a summary of the
findings. With 55 respondents requesting a copy of the
summary, it was decided to send them a copy of the complete
study with the option of purchasing the copy or recycling
it through the author to another agency seeking this kind
of information. About 15 copies were recycled and 29 were
purchased as the result of all mailings of 75 copies.

CONCLUSIONS

In arriving at some conclusions concerning the results of this investigation into the development of the
PACT idea, there are essentially two specific considerations: (1) to what extent is PACT a viable idea in North
America? and (2) what evidence is revealed by this investigation to support the contention that the public access
idea is a potent means of affecting sociocultural change
designed to bring improved social conditions to a community? An extensive statement on the viability of the
PACT idea exists in the following analysis of the data obtained by the PACT questionnaire. There are 29 divisions

of this statement (the question concerning advantages and disadvantages of PACT, though a two-part question, is handled as one division) based on the 30 questions of the survey. Drawn from the analysis of each division are the following statements of conclusion concerning the apparent viability of the idea:

U.S. city government bears more responsibility than Canadian city government in the development of cable television in general and PACT in particular.

The response of the U.S. mayors' offices to the survey questionnaire was 77 percent. The relatively low response (56 percent) from the Canadian cities might be attributed, in part, to the federal licensing of cable systems in Canada, which tends to short-circuit city hall regulatory powers as they are exercised in the United States through the franchising process. An 82 percent response, for example, from the top-50 cities of the United States indicates good potential for the viability of the idea in the larger urban centers.

The amount of activity identified by response to some questions in the questionnaire might appear unimpressive, but in light of the fact that 66 of the 105 responding cities did not even have cable this "miniscule" growth takes on considerably more importance.

At least 12 (more than 10 percent) of the respondent cities were producing PACT programing that involved a sizable assortment of models spread over the three categories of cities. Six of those experiments were in the United States where, with the exceptions of San Diego and Los Angeles, installation of cable television in the 100 largest television markets was an impossible venture until after March 31, 1972. Obtaining FCC permission to build a system is a relatively slow process, especially if local broadcast interests file objections with the FCC concerning the competition of a cable system. The survey shows that very few city officials were involved with the promotion of PACT, although a sizable number were showing some interest in securing information about the idea. It is probable that the natural pressures of the viability of the idea will soon involve a considerable number of city-hall manhours at the executive level. With five cities report-

ing at least two studios dedicated exclusively to PACT
production, a small but important development has been
secured for this very necessary condition to insuring
totally independent community productions. In addition,
the number of portapak videotape units reported to exist
in the cities surveyed is not especially impressive at
first glance, but actually the seeds of growth, seen in
the existence of the necessary technology for the job, are
already widely scattered. Only two of the 45 respondents
to the inquiry concerning the number of portapaks in the
city indicated that there were none. With six cities reporting 51 or more portapak units on hand, the viability
of the PACT idea seems well expressed by them, as representative cities. A valuable model, it would seem, insofar as it possesses potential for "seeding" the PACT idea
through the proliferation of the use of the portapak videotape unit, has been functioning for some months now at the
University of Montreal. According to the director of
audio visual services at the university, students have access to 56 portapak units and may enroll in 30 hours of
coursework involving application of the portable videotape
technology. Simon Fraser University in Burnaby, British
Columbia, has 27 portapak units for checkout but no com-
panion course on the application of the unit. Students
are encouraged to use the portapak both for course projects
and for personal projects and are provided editing facilities on campus that are open 24 hours per day.[4] These
are two models that bear keen attention.

The two active PACT speakers bureaus and the four being planned, as reflected by the questionnaire response,
constitute a very modest beginning for this valuable means
of giving the idea mobility. Along with workshops, speakers bureaus are undoubtedly the most expeditious and effective method of arousing grassroots understanding, and thus
grassroots support, for the public access idea. A model
speakers bureau does not now exist, but the experiment
under way in Kansas City, Missouri, co-sponsored by church
groups and city hall, may manifest a program worthy of
considerable emulation. Speakers bureaus should, in time,
serve as yet another index of PACT viability, especially
during the earlier years of development.

        With most PACT experiments originating from
    privately owned systems (multiple-system owner
    and local owner), the viability of the PACT idea
    is considerably influenced by the social con-
    science and business acumen of cable operators
    engaged in the pursuit of profit.

68

To date, according to the survey, cable owners are
providing a lion's share of the financial support. This
is a very stable source of support, and it definitely can
be viewed as a positive contribution to the viability of
the PACT idea, but, quite obviously, definite steps must
be taken to provide a stable financial base for PACT, including a clearly defined separation of powers between
the cable operator and the community entity charged with
supervising the proceeds of the public access opportunity.
Certainly, it is prudent to believe that the healthy viability of the idea is definitely linked to a financial base
uniformly regulated across each nation. It is equally
prudent to believe that a stifling condition exists if the
viability of the PACT experiments is frequently foreclosed
by anemic funding. The prime motivator for the development
of PACT in the responding communities appears to be the
cable operator, rather than such potential motivators as
city government, citizens' groups, or social-action agencies. The data obtained from the question inquiring after
the number and kind of cable systems in the city dictate
the conclusion that local privately owned and multiple-
owner systems predominate. It could be argued that publicly
owned systems or a combination of publicly and privately
owned systems might exercise more vigor than privately
owned systems alone in assisting the development of PACT.
If a stable financial base is secured for PACT programing
through the suggested arrangement of setting aside a percentage of the gross revenue of each system then the privately owned system, with its deep, abiding loyalty to
the profit motive, might logically generate more financial
support for PACT production than a publicly owned system.
Of the 44 systems reported from the various cities, 42
were primarily in business to return a financial profit.

> City governments are for the most part in
> accord with the federal government dictum that
> public participation in the media is generally
> advantageous to the community.

Most respondents, in answer to the somewhat rhetori-
cal question concerning the advantages of PACT, saw the
greatest advantage to be in the new opportunity for community participation in the cable television production
process. If most city governments acknowledge that this
opportunity is to their advantage then PACT is, indeed,
a viable idea. If there is a common thread that leads
through the various disadvantages listed by the respondents, it would have to do with an anticipated lack of

promotional wisdom and sense of fair play that might sometimes dominate the makeup of citizens committees controlling PACT policy and production. The fact that 29 names
and positions of city officials from 15 major North American cities were volunteered as candidates for PACT information exchange suggests considerable potential for the
validity of the idea in city hall's view.

Neither the federal government's restrictions
nor fear of public abuse of free access is any
longer considered to be among the major blocks to
the development of PACT in the big cities.

There is little indication from the various sources
of information for the study that fear of pornographic,
indecent, and libelous programing will present a barrier
to the viability of the PACT idea. With earlier fears
regarding pornographic, indecent, andlibelous programing
proving to be largely unfounded, those minor problems that
do exist with this aspect of the development of public access are being dealt with routinely. Apathy and lack of
money are the greatest obstacles to the development of
cable television and its accompanying dividends, in the
view of the respondents. Enthusiasm stems from new information, and money invariably is attracted to enthusiasm
for a new enterprise. Federal restrictions imposed on
the new enterprise of cable television have diminished
considerably in the large cities, especially in the United
States; a flourishing industry has resulted, thus demon-
strating an abundance of natural enthusiasm for the new
enterprise. It might very well be that the respondents
have mistaken ignorance for apathy. As might be expected
at this early stage in the development of PACT, city hall
is not being deluged with inquiries concerning public access. There is no good reason for most citizens of the
United States and Canada to be aware of the existence of
the opportunity for public access to the medium of cable
television. Very possibly, especially in the United
States, only after each city has been the home of at
least one thoroughly publicized public-access workshop
can city hall be expected to indeed become one of the
lively sources of PACT information. One of the more reliable growth gauges of the idea should be the call-rate
at city hall for PACT information.

Many cities are now making general plans for
PACT, but it seems that insufficient professional

consultation is being solicited for the very complex task of advanced planning.

Approximately one-third of the cities responding to
the question concerning PACT planning reported doing some
in 1973. Those 14 cities represent a broad base of PACT
growth, should their plans result in community production
over the next few months. It appears from the nature of
the cable planning identified by the 25 respondents answer-
ing this question that the number of cities capable of providing PACT could very well double within the next year.
The concern has been expressed that there is a low level
of investment by city government in professional counsel
with regard to cable planning. This is especially true
of the Canadian cities that responded to the question concerning whether or not they had commissioned a study of
cable television for their city. Considering the tremendous impact cable television will undoubtedly have on the
lives of city dwellers as the variety of cable services
multiplies, it must be expected that city fathers are
morally and ethically obligated to take the most diligent
care possible in the planning and nurturing of this new
scheme for personal and mass communication. The larger
cities in the Category II cities of the United States are
best represented in the list of respondents who indicated
they have sought, or plan to seek, professional counsel
in their planning for cable. Although many sources are
being tapped for information by the city executive offices
in their planning, the Cable Information Center of the
Urban Institute in Washington, D.C., was listed considerably more often than any other source, with the exception
of the National League of Cities. There is no reason not
to believe that for U.S. cities, and to some extent for
Canadian cities, these are excellent sources of help.
It is to be expected that when the cities become more involved with planning for PACT in particular, as distin-
guished from the earlier general planning for cable television, they may find it fruitful to consult with the pioneer public access "seeding" organizations, such as Challenge for Change or New York University's Alternate Media
Center. Such a course would be a most reassuring sign
that the idea was not being corrupted during the uncertain
experimental years. But some cities may, at their peril,
choose to apply scant expertise to the design and operation
of their PACT experiment(s), just as some may choose to
involve the lives of millions of people for years to come
in the natural blunders incumbent in franchises granted

with no more guidance than comes from a part-time cable committee of laymen who are properly overwhelmed by the complexity of their commission, which is received from an equally confused mayor.

There is little indication from the data, for example, that much planning has been conducted in the area of developing special-interest public access networks such as might bring together isolated pockets of a particular ethnic group in a large metropolitan area. Awareness of this possibility is probably one of the least developed of the many opportunities posed by public access. Rockford, Illinois, now serves as one of the lonely models for this facet of the PACT evolution. A substantial number of cities do have plans for installation of the two-way capability for both audio and video within the next five years. Canadian cities, however, are for the most part not involved with planning for this service. This is not to say that Canadian cable television system operators are not planning for this capability. It should be mentioned that although the cable operators were not surveyed directly, several Canadian questionnaire respondents, like some U.S. respondents, were cable operators or managers. In each instance the mayor's office forwarded the questionnaire to the cable operator. PACT models do not seem to play an especially important role--as yet--for those respondents just beginning their public access planning. The burgeoning development of public access workshops around the country should result in a more extensive sharing of information between the novice and the experienced. The data indicate that much of the information reaching city hall about public access has come from cable companies during the franchising process. For those interested in the welfare of PACT, it follows that there should be some merit in publicizing comprehensive franchise models or cable ordinance models that adequately provide for the public access idea.

In response to the request for additional information stated in the survey, 20 respondents provided comments or materials that included copies of franchises or city ordinances, cable reports, personal statements, and personal letters directed to the proposition of community participation in cable communication, which suggests no serious lack of willingness to share information about this new field among city officials and their cable companies. The request of 55 respondents for a copy of the summary of the PACT survey supports the conclusion that many cities are not making general plans for PACT.

In summarizing the data obtained by the questionnaire, it can be concluded that city halls in North America appear to be on the verge of contributing solid support to the proliferation of the PACT idea. There is evidence that considerable cable planning is under way in the big cities; that much of the planning is limited to in-house studies; that counsel, when sought, comes from many sources, but the National League of Cities and the Cable Information Center predominate; and that the idea of public access is a high priority issue with many civic leaders and cable operators both in Canada and in the United States. It is the will of the federal governments of both these nations that public access to cable television be provided, and there is little or no evidence in the response generated by the questionnaire that the wisdom of this will is in dispute.

Information obtained directly from those who have experimented with public access supports the contention that the idea applied is a potent means of affecting socio-cultural change designed to bring improved social conditions to a community.

In the first issue of <u>Newsletter Challenge for Change</u>, published in the spring of 1968, the first popular documentation of the public access idea is discovered. On the opening page of that first issue Hugo McPherson, film commissioner for the Canadian government, reflected on the raison d'etre and potency of the Challenge for Change program.:

When the National Film Board undertook the Challenge for Change program, eighteen months ago, we declared that its objectives were to help eradicate the causes of poverty by provoking basic social change.
Why should a proposal come from the Film Board? The eradication of poverty demands unorthodox ideas, and radical solutions based on them require new concepts of communication. For these purposes, film--used imaginatively and unequivocally--is the best medium. In the first place, unorthodox ideas are much more likely to be accepted if presented in emotional as well as intellectual terms, and film excels in communicating emotions; second many members of the audience to be reached are semi-literate, but can generate

group action. Participation on local levels is
a key element in these proposals.
What has Challenge for Change accomplished?
Even by the coolest judgements the program has
been astonishingly effective.[5]

On the next page of that first issue A. J. Phillips,
director of the Special Planning Secretariat of Ottawa
added further reflection on the impact of this new extension of the National Film Board:

Challenge for Change could become the first
film program to be as well remembered in text-
books of public administration as in festival
awards and certificates of merit.

[. . .]

The important results so far are a few miles
of top-flight film with important things said
well about the new world of Canadians--and an
experiment in collaboration for film sponsorship.
If, as a wise Canadian once said "the process is
the product," the long planning meetings in Montreal and Ottawa may well be justified because
they bring together widely varied interests on
common social problems. They exchange ideas and
frustrations, they break down bureaucratic walls
and open people's doors. Besides, some remarkable
film emerges.[6]

As expected, the first project reported in that first
issue was the Colin Low Fogo Island experiment. The final
statement of this report went as follows: "The measurable
returns will not be in for a long time, if ever. What it
is doing, without question, is to force an area, and eventually, several areas, to become more conscious of their
needs and problems; and a more conscious community is far
better able to anticipate and shape its future."[7]
In Sandra Gwyn's report on the seminar at Memorial
University she described "Fogo Today" through the 40-minute
film shot by Roger Hart of Challenge for Change during the
summer of 1971:

In Hart's film the difference between the
Fogo of 1967 and the Fogo of 1971 is symbolized
by a man called Jim Decker:

Five years ago, Decker was one of the few
fishermen on Fogo able to take advantage of what
[George] McRobie describes as "intermediate technology." For [Colin] Low's cameras, he explained
how he built, virtually single-handed, a new deep
water longliner to replace his inshore trap boat.
Decker now runs Fogo's new cooperative shipyard.
The opening sequence of Hart's film shows him supervising the launching of the fifteenth, sixteenth
and seventeenth longliners the co-op has built.

Other sequences explore other developments:
The Improvement Committee, with representatives
from every community; the new central, all denomi-
national high school being built in the exact center of the Island; the fisheries co-operative; the
plans for a new multi-purpose fish plant.[8]

This was the "same" Fogo Island that four years before was
thoroughly fractionated by ten communities of parochial-
ism, where 60 percent of the people were on welfare. Why
such an abrupt change in the socioeconomic condition of
the community? Gwyn observed: "One question the new
film does not satisfactorily answer--a question that probably can never be entirely answered no matter how sophisti-
cated the evaluation technique--is precisely how much influence the original films had on events." She continued
with these quotations from Colin Low and Dan Roberts:

"It's important to remember," said Low, "that
film and VTR do not themselves effect change.
They can only effect change in the hands of social
animators." Perhaps the best answer was given by
Dan Roberts, Manager of the Producers Co-operative,
in a sequence of the new film:
"It's anyone's guess what would have happened
without the films. I think probably our situation
here now would have been quite different. Certainly I don't think we would have had a shipyard
or even a co-operative formed. I'd say what could
have happened--we might have been all gone."[9]

The key factor that must be clearly understood by
any community access group is that the production
objective must be process rather than product.

A PACT production unit's concentrating on volume production of slick product imitative of commercial television

would defy the law of negentropy. In discussing the Lake St. John project with Gwyn, Louis Portugais warned, "People's first reaction, always, is to make a copy of 'bourgeois' television. Then they begin to move out of this phase. Real community television projects onto the screen the vital preoccupations of an area so that the people can assess them, exchange ideas and initiate changes."[10] As Gwyn suggested, "This type of programming cannot jump on the screen fullblown; it has to be allowed to develop organically."[11] Portugais expressed his conviction that in order for community television to have any social impact there must be structured viewing:

> We would transcribe a CBC broadcast on, say, agriculture, and play it for a group a couple of evenings later. Some would have seen it before, some not. But seeing a broadcast, passively, in your living room is not the same as seeing it with a group of people. Afterward, the group would talk about the broadcast: what is that saying to us? We'd tape the discussion and play it over the community channel a few days later. That would get people talking.[12]

Metro Media went through the early phase of heavy emphasis on producing programs for cable and the "video freak" obsession with hardware. But later, cable was placed in its proper perspective. As a result, "Metro Media became much more concerned with what [Metro Media's Chris] Pinney described as 'community advocacy.' Of its fourteen resource personnel, only three were involved full-time with community journalism (making for local cable) the remainder were community animators."[13] Contrary to the theory espoused by some observers as a result of observing the New York City experience with public access (that the large city is too unwieldy for social animation), Metro Media worked on projects "with more than 40 community organizations; concerned with everything from welfare and housing rights to consumerism and ecology."[14]

The Town Talk program of Thunder Bay, Ontario, which was to serve as a PACT model, ran afoul of several errors that dictated its early demise, but at least three lessons related to the law of negentropy have been salvaged from the wreckage of this experiment:

1. It is absolutely necessary to establish a charter board, representative of the entire community. Town Talk

called for the establishment of a charter board but was
not itself a charter board and hence lacked broad-based
grass roots support.

2. Diversified funding is important. In Thunder Bay
virtually all financial support came from Challenge for
Change.

3. An integrated relationship between the community
organization and the program production teams is needed.
In Thunder Bay the crew was chosen primarily for technical
skills and members lacked training in the techniques of
community development.

Another mistake made in Thunder Bay was that too much
money was poured into the experiment, "making it overblown
and unwieldy."[15] With a large budget there exists the
temptation to bypass "the people" and engage professionals
in the making of slick product that will entertain. Jim
Hyder of Thunder Bay made the point:

> The community channel should not be seen as the
> solution to all problems. The solution still
> lies with the people. The worth of the channel
> lies partly in the fact that this facet of our
> rapidly growing technical society may be used to
> further the purposes of people and community--
> rather than the other way around.[16]

In a brief to the CRTC dated April 1971 the directors
of the Normandin cable access experiment offer these recommendations:

> It is one thing to produce programs for the
> pleasure of producing them, but it is another
> thing to organize programs around the basic con-
> cerns of the community.
> First principle: The community is entirely
> responsible for the production of its programs.
> Second principle: The community itself must
> furnish the human resources and must organize it-
> self financially to ensure its programming.
> Third principle: Programming must be orien-
> ted toward exact community objectives.
> Fourth principle: The operation must involve
> the greatest possible number of citizens. Ideally,
> anyone in the community could participate at any
> level of responsibility in the organization.

Fifth principle: The people charged with the direct operation of community television should be exclusively concerned with the goals of the community, to avoid possible conflicts of interests that could occur.[17]

George Stoney of Alternate Media Center, whose public access experience reaches back to a two-year stint with the Challenge for Change program, recently offered this advice:

The way to avoid disillusionment is to be committed from the outset to paying the price of sticking with a new communications medium until it realizes its full potential. Three things are needed:
1. Cheap, relatively reliable equipment for recording material (e.g. SONY portapak). Our job is to help people make their own communication.
2. Workshops. It is essential to work with others, to share troubles and to build on experience.
3. Feed-back. Unless people get the feeling their message is responded to they'll lose heart. In the first two to four years, nothing is going to happen. "The use of half-inch portable equipment and the playing back of videotapes to groups throughout the community is an essential preparation for an introduction to cable. VTR is an instrument, cable is an extension of it."[18]

Saying that "nothing is going to happen" in the first two to four years does not give consideration to the value of the possible processing that can go on as the people of the community come together in an attempt to communicate solutions to social problems they have in common.

For those who are inclined to experiment with the "mysterious [therapeutic] capability of the media," there is guidance of critical importance in this process analysis provided by Anthony Marcus, acting head of the department of psychiatry at the University of British Columbia:

Confronting himself on camera gradually helps a person develop an internal image of himself. External mirroring by verbal means often creates barriers, regardless of the leader's skill. Most individuals have difficulty communicating their

emotional "hang-ups" but video-taping and the
playback evoke a response on the emotional level.
The simple device of reflecting an image magni-
fies the individual's self image. The emotional
dilemma induced by the gap between the image on
screen and the subjective feeling of the viewers,
produces a crisis in which the person attempts to
bring the two aspects into harmony, thus increas-
ing his self-knowledge. He cannot remain aloof
to himself and he is caught in the conflict between actual conduct and inner fearfulness.

Videotaping pinpoints the failure of the individual to recognize his own problems and difficulties. The end result is that he confronts himself, remaining at the same time less defensive
than when someone else confronts him. Video en-
ables an individual to watch specifically for
himself, a reversal of the normal group situation
where it is the leader who is expected to act as
catalyst most of the time. The camera becomes
catalytic, eliminating the personal battle between leader and participant. As the group helps
people to emerge as social beings, videotape as-
sists them to see themselves as others see them.[19]

What activates the process of people seeing themselves
on videotape? This advice from Dorothy Todd Henaut of
Challenge for Change could be beneficial to city halls of
North America:

Communities need help in conceiving and forming
media associations. They need pointers on the
various possible models, on their advantages and
drawbacks. They need help in recognizing the
resources of their community, in seeing beyond
the static patterns they are used to, in finding
allies in unexpected places.[20]

Public access proponents subscribe to the
macrorevolutionary theory that the quality of
life in a community can be substantially enriched
through decentralization of telecommunications
program control.

When this new theory has been applied, there is no
evidence of any serious negative effects, such as balkani-
zation or the formation of hostile units. Rather, the ap-

plication of the decentralization theory has in most instances resulted in what might be termed "mini-mass communications systems" that provide channels for what Archibald MacLeish would call "palaver." That is, the common people of towns and city communities are provided the opportunity for the first time to engage in parley on a regular basis with persons from other levels of culture and sophistication. In several of the PACT experiments there is evidence that use of these new channels for public expression and dialogue has stimulated citizens to obtain a more positive identification and involvement with the shaping of community character.

The prime source of energy stimulating enlargement of the public access idea in the beginning--the innovative National Film Board personnel --and independent "video freaks" are now being joined by a rapidly swelling contingent of traditional organizations, such as national and local church consortiums, cable system operators, and the FCC.

The church consortiums were especially active in 1973 in promoting public access workshops designed to introduce the PACT idea to community leaders and to provide them with a "hands-on" equipment experience. Numerous cable operators have also been sponsoring workshops, sometimes in cooperation with foundation-supported public access "seeding" centers, and they are providing considerable free channel time, equipment, studio space, and technical expertise. The original impetus for this cable operator largess stems from the March 31, 1972, FCC report and order on telecommunications, but now some cable operators see their provision of free public access service as being good public relations, and public relations is once more coming into vogue with business leaders.

PACT is a secular idea that stimulates a religious fervor and devotion that, when applied to the socioeconomic problems of a community, can elevate the quality of life dramatically.

A growing number of scholars are now putting forth the argument that the single most important "changing need" during the close of this century is the justification of religious institutions. In discussing contemporary religion in the United States with Frank Wetzel of the Associated Press, William Hamilton, dean of the college of arts and letters at Portland State University, responded as follows to the question, "Does society do what it wants to do and then create an ethical framework to justify what it is doing, rather than the other way around?"

> Today the fundamental ethical problems of
> at least the religious and probably the nonreli-
> gious imaginations are going to be set by the
> technological environment.
> This means the church, among others, is no
> longer the obviously necessary institution that
> it was when the ethical problems were simply an-
> swers construed out of a tradition. . . . It
> means the church has real problems to justify
> its existence, unless it's simply to move into a
> new conservativism and new piety, which is now
> happening.[21]

Kenneth Leech, chaplain of St. Augustine's College, Canterbury, made this comment in reviewing John Pollock's biography of Billy Graham:

> We must realize that, contrary to Dr. Graham, the
> Gospel is primarily social, not individual. The
> Gospel is a message for human society, it tells
> of a new age, the Kingdom of God. The individual
> certainly shares in this salvation, but only in
> so far as he is part of the new age, and shares
> the life of God. Secondly, this social salvation
> is not merely for the future but for now, and
> there lies the essence of conflict. For Chris-
> tianity is not at home in a competitive, Mammon-
> oriented society. It is not its natural atmos-
> phere, and it is gasping for breath. It is forced
> into resistance or capitulation. But it is also
> forced into pietism or into a serious attempt to
> analyse, and to offer alternatives to, the pres-
> ent unjust ordering of creation.[22]

Associated Press religion writer George W. Cornell began a recent article with this revelation: "As the Watergate scandal unfolds, moral theologians cite a kind of
'White House religion'--a personalized piety detached from
its social demands--as a factor in the affair."[23] Cornell
quotes the Reverend Gabriel Fackre, of Andover Newton Theo-
logical School in Massachusetts as saying, "White House Re-

ligion seeks the salvation of souls but allows the damnation of society." Cornell's research led him to a cross-section of U.S. ministers, priests, and rabbis:

> Rabbi Balfour Brickner, a Reform Jewish
> scholar, says it [White House religion] stems from
> evangelistic revivalism, which separates religion
> from "affairs of the market place, the courthouse,
> the political arena or the business office. Water-
> gate has shown the fallacy of this attitude. It
> may also restore social action to the churches and
> synagogues of America.

> [. . .]

> The Rev. John B. Coburn, an Episcopal theo-
> logian, says that it seems to him that Watergate
> implies not so much a "constitutional crisis" as
> a "moral crisis" in the country.[24]

An earlier conclusion drawn from the general survey for this study suggested that PACT may very well be a religious idea in the sense of a religious a priori. If public access is a religious a priori then the major conclusion of this study asserts that in the face of "religious and probably nonreligious imaginations" being set by technology, such as portapak cable television, the churches might very well return to social action to help create order (negentropy), to justify their existence. PACT is no less important than that to the common man. It is a way for his community to organize; a way to create social-action negentropy through innovative application of technology to the deepening moral crisis.

It is heartening to note that in the August 31, 1973, edition of the National Association of Educational Broadcasters Newsletter that mouthpiece of noncommercial monologist media devoted about 30 percent of its space to citations from the current Harvey Cox book, The Seduction of the Spirit. Cox's message does not mention the PACT idea specifically, but his thesis is strikingly similar to the major conclusion derived from the PACT study:

> As a Homo sapiens I am an incorrigibly story-
> telling animal. Literally I cannot live without
> a story. But I do not have to search long. A
> substitute is readily supplied. The most powerful
> technologies ever devised churn out signals to
> keep me pliable, immature and weak.

The Christian tradition has produced a set
of stories, images and values that provides a
vantage point by which some judgements can be made
about cultural phenomena. . . . The Bible contains
examples of messages to all kinds of people, and
it sees "God" as the communicator par excellence.
He is the "Word," identified in his very essence
with communication. But it is important to notice how God speaks to man. He does not communicate one way; there is dialogue. . . . We could
even say that man learns something from the life
of Christ about the essential structure of authen-
tic human communication. What he learns is that
communication requires the possibility of response.
Ultimately, self-communication entails vulnerability. The mode of God's self-disclosure to man,
in the life of a man who is abused, rejected and
murdered, is not incidental to the content of that
disclosure. God shows himself to be one who is
willing to risk the most dangerous consequences
of dialogue in order to make himself known. In
the Hebrew scriptures the same paradigm of dialogue, vulnerability and the call for response
can be seen.

With this biblical model in mind it seems
inaccurate to call most of our present mass media
"means of communication." In the light of what
Christianity teaches about the essential ingredi-
ents of communication, they are not means of "communication" at all. They are obstacles to communication and means, mainly, for social control,
propaganda and coerced persuasion.[25]

The newsletter citation concludes with the instruction from
Cox that people must involve themselves considerably more
with critical evaluation of supposed communication and in
doing so, reject that which is false. This theology of
media, envisions Cox, should move the human race toward
increased total communication.

PACT is an intelligent way to mobilize socioeconomic
revolution without violence. Through Challenge for
Change/Societe Nouvelle the federal government of Canada
has been directly involved with social action for the past
seven years. U.S. film maker George Stoney of Alternate
Media Center, who spent two years producing Challenge for
Change social action films, "finds Canadian government officials more mature and secure than the Americans he is

used to dealing with, and believes that Canadians understand the idea of government-sponsored subversion. It is
an intelligent way to mobilize social revolution without
violence," in the opinion of Canada's Patrick Watson.[26]
It is a new way in which people are expressing and experi-
encing the religion of social animation, comparable to the
missionary experience of the Company of Young Canadians or
the ACTION Peace Corps/VISTA volunteers. And no doubt
there are many Americans, such as John P. Roche, professor
of political science at Harvard, who would not be threatened by the "subversive" activities of a PACT community.

> In a healthy democracy the majority and the non-
> conformist depend upon each other, and each sup-
> plies a vital component to the whole. Stability
> is provided by the majority while vitality flows
> from the non-conformist. Consequently, the demo-
> crat protects the rights of the non-conformist
> not merely as an act of decency, but more signifi-
> cantly as an imperative for himself and the whole
> society.[27]

The overriding conclusion of this study has to be that
the only major block to the vitality that flows from the
nonconformist PACT idea is ignorance of that idea's existence. This is most assuredly the condition of the majority today. Therefore, before the majority can rise to
stabilize the participative mystique, it must somehow have
this new vitality described or explained.

RECOMMENDATIONS

Certainly, empirical investigation of communities now
experimenting with public access is needed to identify the
public access idea in relation to the emerging technology
and the changing needs of future communities. In leafing
through the first 11 issues of the Challenge for Change
newsletters, one finds the story of social animation un-
folding almost romantically as the numerous experimental
projects are invariably described in optimistic or noncri-
tical terms. This is a newsletter and not a scientific
journal, of course. However, when the tenth issue is
reached, an article is encountered that deals directly
with the problem of the evaluation of public access. In
this article the National Film Board's Dan Driscoll of
Toronto asks the critical question, "How are our experi-

ments working?" Driscoll calls for "careful project definition, self-criticism, corrective procedures, balance between system design and process integration, all tending towards innovative social research. . . . An obstacle to this is our tendency to link our personal feelings of well-being and achievement with the interpretations we give to results obtained."[28]

The kind of innovative research Driscoll requests appears in a 400-page study of the NFB's community television experiment in the Lake St. John region of Quebec. This study reveals very positive results to have been obtained with the PACT experiment there. A summary of the general study of public access across Canada conducted by the church consortium was originally scheduled for release in 1972 but, apparently, still remains in the making. Actually, most public access experiments in the United States and Canada are not sufficiently mature for objective longitudinal study.

If we use D. Stuart Conger's terminology, the public access idea might best be described as a "social invention." According to Conger, "A social invention is a new law, organization or procedure that changes the ways in which people relate to themselves or to each other, either individually or collectively."[29] Conger became interested in social inventions while directing a Canadian social invention called NewStart, which he described as "a quasi non-governmental organization established jointly by the federal Department of Manpower and Immigration and the Saskatchewan provincial Department of Education."[30] Its purpose was to study new methods of counseling and training adults. Conger calls for the setting up of heuristic energy centers designed to cultivate social invention:

> The methods that are used today to solve social problems are about 4000 years old, whereas the methods used to solve medical, agricultural, transportation and industrial problems are about 25 years old. If we can establish social invention centers, we can create solutions to our age-old social problems, and can rid society of racial strife, mental illness, crime and poverty. This is a goal worth working for.[31]

Challenge for Change/Societe Nouvelle and AMC are examples of model social invention centers in North America. Their leadership appears to be priceless for the time, and it is available for the asking.

NOTES

1. John F. Young, Cybernetics (London: American Elsevier, 1969), pp. 15-16.

2. P. M. Martin, ed., Henry Ford's Sayings (New York: The League-for-a-living, 1923), p. 8.

3. Cable Information 1 (February 1972): 1.

4. Information was obtained in conversations with media personnel during attendance at the Third Canadian Educational Communications Conference, Vancouver, British Columbia, June 11-14, 1973.

5. Hugo McPherson, "A Challenge for NFB," Newsletter Challenge for Change 1 (Spring 1968): 2.

6. A. J. Phillips, "A Challenge for Collaboration," Newsletter Challenge for Change 1 (Spring 1968): 3.

7. Ibid., p. 4.

8. Sandra Gwyn, "Origins of the Fogo Process," Film, Video-Tape and Social Change, p. 6.

9. Ibid., p. 6.

10. Ibid., p. 25.

11. Ibid.

12. Ibid.

13. Ibid., p. 20.

14. Ibid.

15. Ibid., p. 24.

16. Jim Hyder, "Some Recommendations on Community TV," Newsletter Challenge for Change Societe Nouvelle 1 (Winter 1971-72): 13.

17. Ibid. (Spring 1972): 16.

18. Quoted in Cable Information 2 (April 1973): 1, 2.

19. Quoted in Gwyn, op. cit., p. 34.

20. Dorothy Todd Henaut, "A Few Notes on Regional Projects," Access 1 (Autumn 1972): 11.

21. Frank Wetzel, "'Death of God' Author Says Churches Puzzled," Kansas City Star, August 4, 1973, p. 3.

22. Kenneth Leech, "Some Objections to Dr. Billy Graham," London Times, September 8, 1973, p. 14.

23. George W. Cornell, "'Lessons' in Watergate," Kansas City Star, August 4, 1973, p. 3.

24. Ibid.

25. Harvey Cox, The Seduction of the Spirit cited in "Feedback: Theology and Telecommunications," NAEB Newsletter 38 (August 31, 1973): 2.

26. Patrick Watson, "Challenge for Change," Artscanada (April 1970), p. 7.

27. John P. Roche, Shadow and Substance (New York: MacMillan, 1964), pp. 58-59.

28. Dan Driscoll, "Can We Evaluate Challenge for Change," <u>Access</u> 1, no. 1 (Fall 1972): 23.

29. D. Stuart Conger, "Social Inventions," <u>The Futurist</u> 7 (August 1973): 150.

30. Ibid., p. 156.

31. Ibid., pp. 149-58.

PACT CHRONOLOGY

1922   A new form of communication between the character and the viewer, "a participation mystique,"[1] was introduced by Robert Flaherty with his first film, Nanook of the North.

1929   The documentary film was born with John Grierson's Drifters, a film containing the power "to bring the citizen's eye in from the ends of the earth to the story, his own story, of what was happening under his nose."[2]

1938   Grierson was invited to Canada to suggest legislation by means of which all of the government's film activities might be centralized and coordinated.

1939   As a result of Grierson's recommendation, the National Film Act was passed creating the National Film Board (NFB).

1945   According to Forsythe Hardy, editor of Grierson on Documentary, "before the end of the war, the National Film Board was outstripping in enterprise and achievement its equivalent organisations in Britain and the United States."[3]

1951   The Bias of Communication by Canadian historian Harold Innis introduced the theory that mass communications are central to man's historical development.

1964   The concerted and rational use of film as a tool of social animation or an instrument of change first appeared in a French Canadian NFB series produced on agriculture management and development by Fernand Dansereau and Raymond Garceau. In the series they helped " . . . la population locale a prendre une conscience globale impliquant tous les parametres en presence, de la situation dans laquelle elle se trouve."[4]

1965   <u>Understanding Media: The Extensions of Man</u> by
         Canadian author Marshall McLuhan was published. Mc-
         Luhan became the only other student of human history
         "to make the history of mass media central to the
         history of civilization at large."[5]

1966   The NFB spawned Challenge for Change: "A program
         designed to improve communications, create greater un-
         derstanding, promote new ideas, and provoke social
         change."[6]

1967   The technology of film making and the skill of
         film maker Colin Low were placed by the NFB, in coop-
         eration with Newfoundland's Memorial University, at
         the disposal of the Fogo Island People of Newfound-
         land. In the process, training of a university film
         crew was begun. A mediation process was applied us-
         ing spinoff films of additional scenes involving peo-
         ple with decision-making power reacting to preceding
         films, which, in turn, were viewed by those involved
         with the original film. Here was synthesized the
         Challenge for Change approach. Colin Low later be-
         came an executive producer at Challenge for Change.

1968   The Donner Canadian Foundation contributed funds
         to the Fogo Island experiment signaling the beginning
         of a considerable interest being developed by founda-
         tions in the public access idea.

         The first community-operated closed-circuit tele-
         vision channel in the United States began operation
         in Dale City, Virginia, during the month of December.
         (This experiment ceased operation in February 1970
         due to lack of funding.)

         Volume 1, number 1 of <u>Newsletter Challenge for
         Change</u> (now called <u>Access</u>) was published and circu-
         lated free of charge in the spring by the NFB.

         Sony Corporation introduced the first portable
         video camera.

         Paul Ryan, studying under visiting lecturer Marshall
         McLuhan at Fordham University in New Jersey, began
         experimenting with portapak (half-inch video) equip-
         ment.

         Dorothy Todd Henaut and Bonnie Klein of Challenge
         for Change, in cooperation with the Montreal community
         group Comite des Citoyens de Saint-Jacques, introduced

89

the portapak outfit as an effective tool for social animation.

A four-week workshop on "the role of film in community development" was held in St. John's, Newfoundland, under the auspices of the NFB and the Extension Service of Memorial University. "The first workshop of its kind, it was an experiment, exploring the uses of communications technology in community organization."[7]

1969   The Canadian Radio-Television Commission (CRTC) in proposing guidelines for cable systems included in its list of priority services to be offered by the systems, "a channel for community programs."[8]

Raindance Corporation, supported by the New York State Council of the Arts, began its function as an important promoter in print of the PACT idea.

1970   Thunder Bay Community Programs of Thunder Bay, Ontario, originated the concept of a "community board" to oversee the cable production.

1971   New York City's government took the lead in the United States in the development of PACT in granting two cable franchises in Manhattan that required each system to provide two public access cable channels: one channel for immediate programing, the other for projected programing.

A PACT service was instituted in Vancouver by Metro Media, which has gone on to pioneer a multimedia, decentralized communities approach to providing the public with access in a metropolitan setting.

Foundation money for "seeding" PACT appeared in the Eastern states of New Jersey and New York and in Washington, D.C. Substantial sums of money were granted for public access experiments by the Markle Foundation, the Stern Fund, the Schumann Fund, the Kaplan Fund, the America the Beautiful Foundation, and the New York State Council of the Arts.

George Stoney became director of the Alternate Media Center (AMC) at New York University, while Bonnie Klein joined Portable Channel of Rochester, New York, as a social animator doing workshops and holding seminars on the PACT process as it involves integration of all media forms. Both Stoney and Klein

90

had apprenticed with the Challenge for Change program, Stoney as an executive producer and Klein as a film maker.

The CRTC, the government of the province of Quebec, the University of Alberta, the University of Calgary, and a consortium of religious denominations in Canada initiated the first research programs on the structure, range, and impact of PACT.

The Unitarian Universalists in Toronto became the first religious body to undertake regular PACT program production and long-range planning of programs to be directed to a general audience of PACT viewers in the local community.

1972   Cable companies in Vancouver, New York, Pennsylvania, New Jersey, and Florida began providing material support for the PACT idea with contributions of equipment and money. American Television and Communications Corporation, the third largest cable owner, announced that it would provide this support to all of its systems.

The FCC required that all systems in the top-100 television markets provide one public access channel free for a minimum of five minutes of programing on a first-come, first-served basis. New cable systems in these markets are also responsible for providing equipment, studio space, and technical assistance sufficient for production and cablecasting of simple format programs. Established systems are protected from having to provide this service until 1977. However, there does exist a substantial moral obligation for these systems to uncouch as soon as possible.

A survey of the Canadian systems by the CRTC reveals a very healthy response to the PACT idea. Fifty-four systems were doing at least two hours of community programing during the week. Forty produced between two and five hours, and 25 produced more than five hours during the week.[9]

NOTES

1. Frances Hubbard Flaherty, <u>The Odyssey of a Film-Maker: Robert Flaherty's Story</u> (Urbana, Ill.: Beta Phi Mu, 1960), p. 18.

2.   Forsythe Hardy, ed., <u>Grierson on Documentary</u> (London: Collins, 1946), p. 15.

3.   Ibid., p. 151.

4.   <u>Medium Media</u> 1 (Autumn 1971): 4.

5.   James W. Carey, "Harold Adams Innis and Marshall McLuhan," <u>McLuhan Pro or Con</u>, ed. Raymond Rosenthal (Baltimore: Penguin Books, 1968), pp. 270-71.

6.   <u>Newsletter Challenge for Change Societe Nouvelle</u> 1 (Spring 1968): 3.

7.   Ibid., p. 2.

8.   CRTC, <u>Cable Television in Canada</u> (January 1971), p. 19.

9.   CRTC, Broadcast Programmes Branch, "The State of Local Programming" (Summer 1972), p. 11.

METHOD OF INVESTIGATION

In order to discover the rationale of PACT and obtain a historical perspective as well as some tangible evidence of the impact on a community of PACT, it was necessary first to survey the very limited library resources (primarily periodical literature) and then to extend that search by means of correspondence with the builders of PACT in the United States and Canada. It is, to some extent, an evolutionary process, obtaining clues from the more readily available library sources as to those other sources from which additional sources might be developed and/or exhausted. To build a substantial foundation for a comprehensive story of PACT, more than 125 individuals, agencies, and organizations were sent a copy of a one paragraph letter designed to "fish" for general information concerning the development of PACT in this country and Canada. The considerable information generated by the letter is summarized in Chapter 2.

THE SAMPLE

Receiving the survey questionnaire described in Chapter 1 were three general categories of respondents, designated Group I, Group II, and Group III--Group I being the largest U.S. city in each of the top-50 television markets; Group II, the largest U.S. city in each of the next 50 largest television markets; Group III, the 50 largest Canadian cities. There was also a category designated Group IV, which was the total sample of 150 cities. (It should be mentioned that 11 cities or 22 percent of the 50 Canadian cities surveyed are French Canadian, necessitating the preparation of a questionnaire in the French language.) The list of the 100 largest television markets was obtained from Broadcasting Cable Sourcebook 1972-73. The 50 largest Canadian cities were obtained from 1972 Canada Year Book, Statistics Canada, which is compiled from the 1971 census.

THE PACT QUESTIONNAIRE

The primary instrument, the questionnaire, which was used to survey the city halls of North America's larger cities, is composed of questions directed, for example, at identifying which city officials, and how many, are actively engaged in promoting public access. There are many people who should want to know how far along the cities are in this work. The survey inquires into the kind of planning these cities have done to date. There are questions about plans presently being formulated and about long-term planning: the type of cable system ownership extant or planned; the number of public access production studios; the availability of financial support for PACT; the availability of PACT equipment and PACT network design; the methods of promoting the idea of PACT in the city and beyond into the greater metropolitan area. In essence, the questionnaire inquires about the success of big city government in North America in meeting the PACT idea.

The purpose of this questionnaire is to determine the status of public access cable television (PACT) in the major television markets of the United States and Canada. For the sake of brevity the acronym PACT has been originated for this study. For the purposes of this study PACT is defined as community cable television programming produced and controlled by a committee of community representatives or by an independent individual in the community. PACT does not include that locally originated programming produced and controlled by the management of the cable system(s).

1. Indicate the approximate number of hours of PACT programming produced each week in your metropolitan area and the number of cable systems involved.

   0 hours _____ by _____ cable operator(s)
   1-2 hours _____ by _____ cable operator(s)
   3-5 hours _____ by _____ cable operator(s)
   6-10 hours _____ by _____ cable operator(s)
   Over 10 hours _____ by _____ cable operator(s)
   Have no idea _____

2. If PACT is being provided in your metropolitan area, what prompted development of this service?

94

3. Indicate below how many separate cable systems of each type of ownership exist in your metropolitan area.

None	_____
Local private	_____
Local public	_____
Multiple systems owner private	_____
Combination of local private and local public	_____
Combination of multiple owner and local public	_____
Other (please describe)	_____

4. Indicate how many of each type of ownership existing in your metropolitan area cablecast PACT.

None	_____
Local private	_____
Local public	_____
Multiple systems owner private	_____
Combination of local private and local public	_____
Combination of multiple owner and local public	_____
Other (please describe)	_____

5. If PACT programming is produced in your city in 1973, please indicate the approximate percentage of financial support it will receive from any of the following sources.

Outside government grant(s)	_____%
Local government grant(s)	_____%
Outside private grant(s)	_____%
Local private grant(s)	_____%
Cable company grant(s)	_____%
Other	_____%

6. Approximately how many inquiries about PACT have been received at your city's executive offices during the past 6 months?

    Local _____   Outside _____

7. Does your metropolitan area have any major plans for PACT in 1973?

    Yes _____   No _____

    If yes, what are they?

8. How many employees of your city government do you estimate are directly involved with aggressive promotion of PACT in your metropolitan area through committee planning, speaking engagements, consultations, etc.? _____

9. To help bring together informants who may be valuable information and communication resources to each other, especially in community programming development, please list the names and positions of those city employees most involved with the promotion of PACT in your city.

   Name_____   Position_____
        _____   _____
        _____   _____
        _____   _____

   May these names be included in the summary of this survey?

   Yes _____   No _____

10. To your knowledge how many production studios available exclusively to PACT are now serving the greater metropolitan area?

    0 ___ 1 ___ 2 ___ 3 ___ 4 ___ 5 ___ 6 ___ 7 ___ more _____

11. What have been the major sources of information for your city's executive office concerning guidelines for developing PACT?

12. How many portable video tape recording units (such as Sony portapaks) do you estimate are now on hand in your city? (Units owned, for example, by the public schools, community colleges, large corporations, and governmental agencies.)

    0   _____
    1-5   _____
    6-10   _____
    11-20   _____
    21-50   _____
    51 or more   _____
    Have no idea   _____

13. What plans does your city have for future development of special interest (that is, for example, ethnic groups, public service organizations, and community information groups) network links in your metropolitan area and beyond?

14. It has been suggested that a community programming board might be the proper body to guard against pornography and libel on public access channels.

    Has your city dealt with this problem?

    Yes _____   No _____

    If yes, how?

15. Has your city ever commissioned a study of cable television in the city and/or the metropolitan area?

    Yes _____   No _____

    If yes, how extensive was the study?

16. Does your city have plans for two-way cable with audio-video dialogue capability?

    Yes _____   No _____

    If yes, how soon?

    1-5 years   _____
    6-10 years   _____
    11-15 years   _____
    16-20 years   _____

17. Has your city established a speakers bureau for the purpose of extending information and obtaining feedback on the local level in the development of a comprehensive metropolitan cable television service?

    Yes _____   No _____

    If no, is a bureau planned? Yes _____   No _____

    What other means of promotion and feedback will be utilized in 1973?

18. What appear to be the major obstacles to development of PACT in your metropolitan area?

Greatest obstacle _____

_____

Second greatest _____

_____

Third greatest _____

_____

19. What appear to be the major advantages to be realized from PACT in your community?

Greatest advantage _____

_____

Second greatest _____

_____

Third greatest _____

_____

Disadvantage _____

_____

20. If members of the executive staff of your city have been studying the development of PACT, which of the communities with PACT that have come to their attention seem to offer the best models for their planning?

Why?

21. Where did the idea or concept of public access (community) television originate?

22. Do you wish to have a copy of the summary that will be prepared of this international PACT survey?

Yes ____   No ____

Your name is _____   Title _____

Please enclose any additional information that you feel would contribute to this study of public access (community) cable television in the United States and Canada.

OBJECTIVES OF THE PACT
QUESTIONNAIRE

As stated in the PACT questionnaire, the questionnaire's purpose, "is to determine the status of public
access cable television (PACT) in the major television
markets of the United States and Canada." The questionnaire is designed to obtain some facts, the analysis of
which should offer information that provides some insight
into the question, "Which of 150 major North American
cities are engaged in developing PACT and roughly how far
along are they in this work?" In his opening address to
the Fourteenth Nobel Symposium in Stockholm in 1969 Arne
Tiselius, head of the Nobel Institute of the Royal Academy
of Sciences, made these comments concerning facts in defer-
ence to the symposium's theme, "The Place of Value in a
World of Facts":

> I referred to the abundance of facts--but
> is there an abundance of _true_ facts? In science
> and technology--yes. In some other fields of
> research--perhaps. But the rest, the facts which
> influence our daily decisions, our valuations,
> our political attitudes? Facts can be and are
> distorted, people are purposely excluded from
> certain facts which may influence their attitudes.
> Distortion of facts in mass communication media
> are tragically one of the most efficient means of
> ruling the world to-day. By mixing true and false
> one can make a terrible brew to manipulate the
> minds of the uncritical, and not only of these.
> Especially to a scientist--like myself--
> such a situation is unbearable. In scientific
> research there is a clear distinction between
> false and true. Any attempt to manipulate this
> is deemed to be a failure. True, wishful think-
> ing or, even worse, human weaknesses may often
> lead us astray. But if we insist, Nature will
> sooner or later strike back--and it hits hard.
> The experiment will answer yes or no, whatever we
> hope for or wish. And if Nature answers "per-
> haps" it just means that we shall have to sit
> down again to think more.[1]

LIMITATIONS OF THE STUDY

Very often a television market in the U.S. survey contains more than one city. The number-one market, for example, contains Linden and Paterson, New Jersey, as well as New York City. In those instances where multiple city markets exist the questionnaire was mailed to the office of the mayor of the largest city in that market. The survey of the 100 largest U.S. television markets, then, includes more than the 100 largest U.S. cities. This limitation did not apply in the survey of the Canadian cities, where television markets that include more than one city have not been established. Rather, mailing the questionnaire to the office of the mayor in the 50 largest cities was identical to surveying the 50 largest Canadian television markets. Because this study was designed to measure PACT activity and its potential development in the larger metropolitan areas of the United States and Canada, it has the limitation of not surveying such important PACT experiments as exist, for example, in Reading, Pennsylvania, and the Lake St. John region of Quebec Province.

The survey was conducted approximately one year after the FCC rules went into effect (March 31, 1972), lifting the long-imposed freeze on the expansion of cable television into the 100 largest U.S. television markets. Systems in existence were given five years to comply with the public access requirement contained in these new rules. As reported in the Kansas City _Times_, "Impact of the rules will not be felt immediately because communities usually take a long time to grant cable franchises, and follow-up FCC clearances take about 18 months."[2] The CRTC first placed particular emphasis on the opportunity for cable licenses to provide public access in its announcements of May 1969 and April 1970.[3]

One final limitation of the study considered here is the difference between the Canadian and U.S. samples in regard to types of regulation. Although federal, state, and municipal control obtains in U.S. cable systems, local control in the form of city ordinances and franchises is the predominant factor shaping the quality and quantity of services rendered. Cable systems in Canada are licensed by the federal government, not franchised by a local entity of government. As a result, Canadian city or municipal governments do not have the natural involvement in cable development or the incumbent responsibility of providing public access incubation that exists with the major U.S. cities. Although an attempt was made to design a universal survey instrument, in some instances the questionnaire is more applicable to the function of the U.S. rather than the Canadian city hall.

In his analysis of probability John Locke provided these guidelines, which can be used to summarize the major limitations of this study:

> Probability then, being to supply the defect of our knowledge and to guide us where it fails, is always conversant about propositions whereof we have no certainty, but only some inducements to receive them for true. The grounds of it are, in short, these two following:
>
> First, the conformity of anything with our own knowledge, observation, and experience.
>
> Secondly, the testimony of others, vouching their observation and experience. In the testimony of others, is to be considered: 1. The number. 2. The integrity. 3. The skill of the witnesses. 4. The design of the author, where it is a testimony out of a book cited. 5. The consistency of the parts, and circumstances of the relation. 6. Contrary testimonies.[4]

In evaluating the responses to the questionnaire, "our own knowledge, observation, and experience" in relationship to the degree of interchange of PACT atoms and molecules in each city is invariably nonexistent. We proceed with such general information as can be garnered from reviewing the literature. The evaluation depends, then, almost entirely on "the testimony of others"--these others being the cities' cable representatives. "The number" of communities surveyed is sufficiently large, but one limitation in the survey sample is the fact that the community television experiment conducted in the Lake Saint John area of Quebec province and the Berks Cable Company public access experiment in Reading, Pennsylvania, are not included. As for "the integrity" of the respondents and their "skill," it is assumed these qualities are equal to the level of their responsibility as suggested by the title of their positions in city government. "The consistency of the parts" can, to some extent, be evaluated in the consistency of the level of local PACT vigor reflected throughout the questionnaire responses. The "circumstances of the relation" variable can be applied to the titles or positions held by those employees indicated to be most involved

with the promotion of PACT. As to "contrary testimonies,"
it is not practical to solicit these for this study.

NOTES

1. Arne Tiselius and Sam Nilsson, ed., The Place of
Value in a World of Facts (New York: Wiley, 1970), p. 13.
2. Kansas City Times, April 1, 1972, p. 9A.
3. CRTC, The Integration of Cable Television in the
Canadian Broadcasting System, a statement, February 26,
1971, in preparation for a public hearing held April 26,
1971, in Montreal, Quebec, p. 26.
4. John Locke, An Essay Concerning Human Understanding, Vol. II, ed. A. C. Fraser (New York: Dover Publications, 1959), pp. 365-66.

SOURCES CONSULTED

Books

Fabun, Don. The Dynamics of Change. Englewood Cliffs, N.J.: Prentice-Hall, 1967.

Flaherty, Frances Hubbard. The Odyssey of a Film-Maker: Robert Flaherty's Story. Urbana, Ill.: Beta Phi Mu, 1960.

Hardy, Forsythe, ed. Grierson on Documentary. London: Collins, 1964.

Heider, Fritz. The Psychology of Interpersonal Relations. New York: Wiley, 1958.

Locke, John. An Essay Concerning Human Understanding. Volume II. Edited by A. C. Fraser. New York: Dover Publications, 1959.

Newman, John Henry. An Essay on the Development of Christian Doctrine. Edited by Charles Frederick Harrold. New York: Longmans, Green, 1949.

Shamberg, Michael, and Raindance Corporation. Guerrilla Television. New York: Holt, Rinehart and Winston, 1971.

Singh, Jagsit. Great Ideas in Information Theory, Language and Cybernetics. New York: Dover, 1966.

Stroud, William. Selected Bibliography on Telecommunications (Cable Systems). Madison: Wisconsin Library Association, 1972.

Young, John F. Cybernetics. London: American Elsevier, 1969.

Periodicals and Journals

Alexander, Herbert E. "Communications and Politics: The Media and the Message." Law and Contemporary Problems 44 (Spring 1969): 255.

Belt, Forest H. "Cable TV--Where It Is and Where It's Going." Popular Electronics, January 1971, p. 26.

Broadcasting. "Open Access: What Happens?" May 1, 1972, p. 47.

_____. "Open Access: What Happens?" May 1, 1973, p. 46.

_____. "Cable Briefs," May 14, 1973, p. 51.

_____. "Teleprompter, NCT a vote for proposal by Open Channel," May 28, 1973, p. 51.

_____. "FCC Access for Cable's Access People," July 9, 1973, pp. 32-33.

Guichard, Marie-Therese. "Le cable cassera le monople." Le Point, no. 11 (December 4, 1972), p. 35.

Knox, William T. "Cable Television." Scientific American, October 1971, pp. 24-25.

Korman, Frank. "Innovations in Telecommunications Technology: A Look Ahead." Educational Broadcasting Review, vol. 6 (October 1972).

Molenda, Michael H. "The Educational Implications of Cable Television (CATV) and Video Cassettes: An Annotated Bibliography." Audiovisual Instruction, vol. 17 (April 1972).

Parker, Herbert L. "Piling Higher and Deeper; The Shame of the Ph.D." Change 2 (November-December 1970): 53.

Reports, Statements, Speeches,
and Conferences

Canadian Radio-Television Commission, Broadcast Programmes Branch. "The State of Local Programming," Summer 1972.

Canadian Radio-Television Commission. Canadian Broadcasting: A Single System. A policy statement on cable television, July 16, 1971.

_____. The Integration of Cable Television in the Canadian Broadcasting System. A statement, February 26, 1971, in preparation for the public hearing held in Montreal, April 26, 1971.

_____. Canadian Television in Canada, January 1971.

Center for the Analysis of Public Issues. Public Access Channels: The New York Experience. A report for the City of New York, March 1972.

_____. New York CATV Supplement, supplement no. 2, August 1971.

_____. Public Issues, supplement no. 1, July 1971.

Changing Communications. A national college conference on cable television sponsored by the National Cable Television Association, George Washington University, February 11, 1972.

Feldman, N. E. Cable Television: Opportunities and Problems in Local Program Origination. A Rand report for the Ford Foundation, September 1970.

Fieland, Shelley. "Questions and Answers about Open Channel." An Open Channel letter, July 12, 1972.

Gwyn, Sandra. Film, Video-Tape and Social Change. A report on a seminar organized by the Extension Service, Memorial University of Newfoundland, St. John's, Newfoundland, March 13-25, 1972.

League of Oregon Cities. "Cable Distribution of Television: A Survey of Practises in 104 Oregon Cities." Policy and Practice in Oregon Cities. Bureau of Municipal Research and Service, University of Oregon, June 1967.

MITRE Corporation. Urban Cable Systems. A report, May 1972.

Murphy, Arthur R., Jr. "Communications: Mass Without Meaning." Vital Speeches of the Day, vol. 44 (1967).

National Cable Television Association. Guidelines for Access: A Report by NCTA, August 1972.

Schlafly, Hubert J. The Real World of Technological Evolution in Broadband Communication. A report prepared for the Sloan Commission on Communications, September 1970.

Smith, Stratford E. "The Emergence of CATV: A Look at the Evolution of a Revolution." Proceedings of the IEEE 57 (July 1970): 968.

Tate, Charles, ed. Cablevision in the Cities. New York: Harper & Row, 1965.

Thayer, Lee. "On Human Communication and Social Development." Economies et Societes: La Communication II, volume V. Geneva: Librairie Droz, 1971.

Newsletters

Arlen, Ann. "Will Public Access Be the Second Coming of Television?" Foundation News, May-June 1972.

Cable Information, vol. 1 (July 1972).

_____, vol. 2 (April 1973).

_____, vol. 2 (June 1973).

Dansereau, Fernand. "Channeling Change in Quebec: Fernand Dansereau's Saint-Jerome." Newsletter Challenge for Change 1 (Fall 1968): 10.

Driscoll, Dan. Can We Evaluate Challenge for Change?" Access 1 (Fall 1972): 23.

Grierson, John. "Memo to Michelle about Decentralizing the Means of Production." Newsletter Challenge for Change Societe Nouvelle (Spring 1972): 5.

Hannah, Chad. "Thunder Bay Community Programs." Newsletter Challenge for Change Societe Nouvelle 1 (Spring 1972): 6-9.

Henaut, Dorothy Todd. "Galloping Videoitis." Newsletter Challenge for Change Societe Nouvelle 1 (Spring 1972): 3.

_____, and Bonnie Klein. "In the Hands of Citizens: A Video Report." Newsletter Challenge for Change 1 (Spring-Summer 1969): 2-3.

Medium Media 1 (Autumn 1971): 4.

Pinney, Chris. Metro Media Print-Out, vol. 1 (July 1972).

Prinn, Elizabeth. Newsletter Challenge for Change Societe Nouvelle, vol. 1 (Spring 1972).

Unitarian Universalist World, vol. 2. "Canadian UUS to Study TV," June 1, 1971, p. 1.

_____, vol. 3. "Community TV in Canada," February 15, 1972, p. 1.

Newspaper Articles

Cornell, George W. "'Lessons' in Watergate." Kansas City Star, August 4, 1973, p. 3.

Edmonton Journal. "CRTC Pushes Local Talent," July 4, 1973, p. 68.

Evening Outlook (Santa Monica, Calif.). "Cable Television Lures Churches," June 17, 1972, p. 10.

Gent, George. "Public Access TV Here Undergoing Growing Pains." New York Times, October 26, 1971, p. 83.

Gould, Jack. "Waterbury Officials Planning for First CATV City." New York Times, January 20, 1972.

Jersey Journal and Jersey Observer (New Jersey). "Hoboken Will Open Videotape Workshop," June 1972, p. 1.

Kansas City Times. "FCC Allows Cable TV to Expand Broadcasting," April 1, 1972, p. 9A.

Leech, Kenneth. "Some Objections to Dr. Billy Graham." London Times, September 8, 1973, p. 14.

O'Connor, John J. "Public Access Experiments on Cable TV Advancing." New York Times, June 6, 1972, p. 82.

_____. New York Times, June 6, 1972, p. 82.

Wagner, Joyce. "An American Family Upset by Monster Image." Kansas City Star, February 25, 1973, p. 43.

Walz, Jay. "Toronto Station Swings with X-Rated Movies." Kansas City Times, March 15, 1973, p. 8A.

Wetzel, Frank. "'Death of God' Author Says Churches Puzzled." Kansas City Star, August 4, 1973.

PRIMARY AND SECONDARY U.S. SOURCES

The Federal Communications Commission (FCC)

The Artist as Politician and the Politician as Artist.
Remarks by Commissioner Nicholas Johnson, FCC, prepared
for delivery at the Annual Conference of the Associated
Council of the Arts, Washington, D.C., May 24, 1971.

> The potential of cable television, which can
> be used with 1/2-inch video tape for a relatively
> inexpensive, do-it-yourself-television system,
> has excited a growing number of young artists ac-
> ross the country. These artists see television
> as a virtually untapped medium of creative ex-
> pression. Many find themselves involved in poli-
> tics simply because their chosen artistic medium
> is presently controlled by antagonistic forces.

Cable Television Rules Statements by Commissioners. Issued by members of the FCC in conjunction with the Report
and Order on Cable Television adopted February 2, 1972.
Opinion of Commissioner Nicholas Johnson, concurring in
part and dissenting in part.

> Cable Development: A Model
> The model I have outlined ought to have the
> support of most people of independent mind . . .
> "free entrepreneurs" and "regulators" alike. It
> serves the "public interest" and is wholly consis-
> tent with the profit motive. The problem, of
> course, is that it does not have the support of
> the most powerful broadcasters--a group whose po-
> litical influence is unrivaled in our time.

CATV: New Hope for the Minorities. Remarks by Commis-
sioner Nicholas Johnson, FCC, prepared for delivery at the
Urban CATV workshop, "Minority Business Opportunities in
Cable Television," sponsored by the Urban Institute, Ur-
ban Communications Group, and Black Efforts for Soul in
Television, Washington, D.C., June 26, 1971.

The great promise of cable is that it offers
a solution to the three main problems that plague
our current closed-over-the-air broadcast system:
(1) poor reception
(2) lack of program diversity; and
(3) too few outlets for truly local expression.
Cable's prime potential is to help turn our
depersonalized cities into functioning neighborhoods and local communities.

Community Antenna Television (CATV). FCC, information
bulletin, November 1970.

On June 23, 1970, in addition to adopting
rules on CATV systems ownership, the Commission
proposed a series of new rules designed to advance the potential of CATV as a communications
medium. . . . The Commission also asked for comments on requirements that CATV systems, in addition to a channel for program origination, provide a channel for use without charge by local
governments and for free political broadcasts
during primary and general elections; a public
access channel to permit local citizens and
groups to present views on issues with which
they are concerned; a leased channel, which would
be available for commercial operation by third parties; and channels to be used expressly for educational purposes.

Memorandum Opinion and Order on Reconsideration of the
Cable Television Report and Order Adopted June 16, 1972,
released June 26, 1972, in response to the reaction generated by the Cable Television Report and Order Adopted February 2, 1972, by FCC.

We doubt very much if, in new systems in
major markets, a scarcity of access channels will
arise from a cable operator's excessive use of
bandwidth for his own origination purposes; but
if a problem should arise, we shall be alert to
take action to maintain our emphasis on the provision of access channels.
For each additional broadcast signal carried,
such a system will have to provide one access channel. The first additional signal will be comple-
mented by a public access channel, the second by
an educational access channel, the third by a gov-

ernmental access channel, and any others by leased channels.

Books

Jencks, Charles. <u>Architecture 2000</u>. New York: Praeger, 1971.

In general systems theory the machine, nature and culture are all just different levels of organized system working in opposition to the trend towards entropy or disorganization. This opposition may emerge as the most fundamental polarity underlying man's belief and experience. The general trend of matter toward its most dispersed, disorganized state--the ultimate quiescence, monotony and randomness which the Second Law of Thermodynamics postulates as the outcome of the universe--is countered by the tendency for the three "privileged" systems to increase in their order and complexity. One might imagine a morality based on this distinction which identified the good with those systems which are continually self-transcendent and the bad with those that become more and more simple, monotonous and disorganized. Thus the nature/culture/machine distinction would be replaced by the more organized/less organized distinction and complexity would, <u>ipso facto</u>, be better than simplicity.

Frederiksen, H. Allan. <u>Community Access Video</u>. Menlo Park, Calif.: Nowels, 1972.

The alternate counter-culture video movement is happening. There are few sources of information on how to become a participant in experimental video. One function of this book is to provide information on how to participate in the process of alternate video and survive.
The necessary video equipment and means of getting alternate video information to large numbers of people are already available to you in your community. You simply have to know where to look and methods of gaining access to hardware and distribution systems. This book is intended to provide you with that information.

110

Price, Monroe E., and John Wicklein. Cable Television: A Guide for Citizen Action. Philadelphia: United Church Press, 1972.

Whether a community corporation or a private operator--or a combination of both--gets the franchise, your group can press for establishment of an officially recognized watchdog agency to protect the public's interest in the system. One way to do this would be to set up a Community Cable Television Board.

This community board would be charged with seeing to it that the cable operator's public-service promises are carried out, and that free and nondiscriminating access to public channels is guaranteed to people in the community. Even unpopular people!

The board should be chartered to the municipality, so it has independent legal authority. But it must be divorced from the municipal government. Funds to operate it might come from a fee paid directly by the cable operator to the Board. Board membership should be made up of a wide cross section of community interest groups and factions, both organized and unorganized.

Shamberg, Michael, and the Raindance Corporation. Guerrilla Television. New York: Holt, Rinehart and Winston, 1971.

Broadcast television is structurally unsound. . . . Reforming broadcast television would be, as Frank Gillette says, "like building a healthy dinosaur." Healthy systems share the following characteristics:
1. they support a high variety of forms, or diversity rather than uniformity;
2. they are complex, not simple;
3. they minimize redundancy and are thus negenthropic;
4. they are symbiotic rather than competitive;
5. they trend toward decentralization and heterogeneity; and
6. they are stable as a result of the above.
Under these ecological rules, broadcast television becomes beast television.

Tate, Charles, ed. <u>Cable Television in the Cities:</u> <u>Commu-</u>
<u>nity Control, Public Access and Minority Ownership</u>. Wash-
ington, D.C.: The Urban Institute, 1971.

The smaller the definition of the market area,
the better off you are". . . . In a city, the cost
of constructing a cable company will be far greater
than the cost in rural areas, but since there are
more people concentrated in a city, it is possible
to have a larger number of customers within a
small section of the city. Mr. Christensen, as
well as later conference panelists and speakers,
opposed plans to issue only one franchise for an
entire city. [Gary] Christensen [General Counsel,
National Cable Television Association], questioned
by [Ted] Ledbetter [President, Urban Communications
Group], on what position his organization took on
minority ownership, said he could not promise an
NCTA policy supporting community control or minority ownership, but his own belief is that "cable
should be owned by the people in the community
where it exists."

He added, "If you can't own it, you should
go into partnership. If you can't go into partnership, you should have access to at least one
channel for your own communication needs."

Newsletters

Alternate Media Center. <u>A Story about People</u>, April 1972.

AMC was thoroughly convinced after seeing
the city and meeting the Berks and ATC staffs that
Reading [Pennsylvania] was the place to conduct
the project that is currently, according to Mrs.
Burns "without precedent in the rapidly growing
cable television industry."

Mrs. [Phyllis] Johnson moved to Reading on a
temporary basis in mid-January. Her function has
been to organize the Community Video Workshop,
teach classes on operating the Sony half-inch
equipment, and work the Berks TV Cable closed-
circuit department in scheduling. A twelve-inch
newspaper ad and a news story run in the <u>Reading</u>
<u>Sunday Eagle</u> January 23rd ignited the spark that
exploded into an inferno of enthusiasm throughout
the community.

112

Classes have been running regularly since
that initial publicity was released. The group
has exceeded fifty members and is still growing.
Over two hundred tapes including practice sessions,
workshop meetings and classes have been cut. The
signout sheets are evidence of the frenzy of interest and dedication to this project.

American Association of School Administrators. "Cable TV
Conference Attracts 250." School Administrator. Comments
by editor Martha A. Gable, June 1972.

A groundswell of citizen interest in the possibilities of cable TV for public service was dem-
onstrated by the 250 representatives of school districts, colleges, religious groups, urban planners,
city councils, ethnic groups and many public and
private nonprofit organizations who came from
coast to coast [at their own expense] to a national
cable TV conference held in Washington, D.C., May
5-6. Publicable, a coalition of individuals representing nearly 100 public service organizations,
planned the conference to inform its constituents
of their responsibilities in assuring public service provisions when local franchises are granted.

Arlen, Ann. "Will Public Access Be the Second Coming of
Television?" Foundation News, May-June 1972.

With few exceptions, people involved with
program production for Public Access receive little or no pay. They are a dedicated lot, and
many have had philanthropic support. Open Channel, organized by Thea Sklover to provide taping
facilities and personnel to groups wishing to
put programming onto the public channels, got
started with a $19,000 grant from the John and
Mary R. Markle Foundation and $15,000 grant from
the Stern Fund. Open Channel has taped programs
for more than 80 organizations and has more than
that waiting. They have also done some of the
most ambitious public-channel programming, including a two-and-one-half hour music "special" from
a Harlem church.

Association for Supervision and Curriculum Development,
NEA. "Cable TV--Protecting Its Future in Education." Interpretations, November 1971.

Even if the business is going to be very
profitable in the long run the immediate prospect may be one of high expenditures and uncertain returns. Still, the moral of this story is
that something of enormous potential value is on
the auction block and before they get that franchise the bidders may be willing to go pretty
high. It is time to stop depending on "gentlemen's
agreements" and to go for contracts that protect
the public and the schools.

Clearinghouse on Educational Media and Technology. ERIC
at Stanford, special issue, December 1971.

Guerrilla Television is what author Michael
Shamberg calls a "print-out" of the collective
experiences of Raindance and other video groups
in experimenting with new media forms, together
with a general scheme for decentralizing television. A multitude of suggestions for using video
and designing video systems in the contexts of
community, education, and self-exploration are
prefaced by a lengthy analysis of media systems
as ecological forms. The prime focus is toward
developing a sense of "media ecology," i.e. understanding the psychic consequences of established information forms like broadcast television and print-based learning in the context of
personal and social evolution--and perceiving the
potential advantages of newer forms, such as the
feedback capabilities of 1/2" video.

Feedback: Feedforward. A newsletter introducing Portable
Channel, Rochester, N.Y., March 1972.

In addition to operating the Equipment Pool,
we plan to do much more ALTERNATE TELEVISION . . .
making tapes with information about our community
that we don't see on regular TV or read in our
newspapers. We need and want Your ideas, invita-
tions, scoops, etc. about news--in its broadest
sense--that you want to know about or want others
to know about. About the Advisory Committee we
mentioned in the last FEEDBACK: FEEDFORWARD;
we've decided against a formal committee with
"professional board members" and regular meetings for the time being. We have developed instead, a group of 20 or so people* with whom we

114

consult as the need arises, keeping open the possibility of formalizing the relationship sometime in the future. *(There's a list available on request.)

National Association of Educational Broadcasters. <u>Newsletter</u>. February 21, 1972.

If record enrollment and enthusiastic reaction at the first Educational Broadcasting Institute on Telecommunications Center Planning are indicative, interest in harnessing the potential of new communications technology is high.

Associate director of Professional Services George Hall described response to it as "impressive and exciting. . . . Every metropolitan, urban and rural area in our states and territories should have adequately staffed and equipped facilities for planning . . . designing . . . producing . . . distributing and evaluating a wide range of vitally needed television and radio materials. Employing broadcasting . . . cable . . . ITFS . . . and the new audio-video cartridge technologies . . . these locally responsible public telecommunications complexes should actively assist our society in a number of significant directions.

_____, April 24, 1972.

In remarks delivered at the opening session of the 1972 Public Television Conference, "Toward '77 and Beyond," <u>NAEB President William Harley announced plans to form four NAEB "Working Parties" that would tackle problems related to transforming educational radio and television stations into Telecommunications Centers</u>.

Such multi-modal centers, first proposed by Mr. Harley last May, would relate modern communications technologies--principally cable television, but also cartridge and disk systems, cassettes and other media--to the total range of educational and public service needs.

"We need to move toward a projection of dollars and facilities which will bring the resources and capacities of communication technology into a position where they will become an integral means of education, necessary instruments for social dis-

course, and the very embodiment of a communication process in which many people have access to
the media, as well as the media having access to
the people.

_____, July 31, 1972.

In declining to sign the hard-won funding
legislation Mr. Nixon said the measure "ignores
some serious questions which must be resolved
before any long-range public broadcasting financing can be soundly devised. . . . By not placing
adequate emphasis on localism, HR13918 threatens
to erode substantially public broadcasting's impressive potential for promoting innovative and
diverse cultural and educational programming.
. . . I ask the Congress to follow my budget
recommendation by enacting a one-year extension
of the Corporation's authorization and providing
it $45 million.

Open Channel. "Questions and Answers about Open Channel."
Newsletter, July 1972.

Open Channel is an independent, non-profit
corporation created to assure effective public access to cable television. Headquartered in New
York City, it provides information, production
assistance, training and the general support
needed to make public access TV a reality for the
community.
Open Channel was established in October, 1971
under the direction of Theodora Sklover, a communications specialist and educator who has partici-
pated actively in formulating the concept of public access. A professional talent pool of over
200 television and film producers, directors,
writers, cameramen, audio and lighting engineers
volunteer their time and expertise to Open Channel productions.

Please Read Me. An open letter from Carol and Maxi, the
producers of "Are You There?" Cape May, N.J., Winter 1972.

In the beginning we had the idea of experimenting with 1/2" videotape to create an alternate means of community communications in Cape May
County. In July 1971 we received a $500 grant from
the America the Beautiful Fund, 1/2" videotape

equipment from the Alternate Media Center of New
York University, and a home and transportation
from Mr. and Mrs. Joseph Cohen. . . . We began
to spread our enthusiasm and involve individuals
and local organizations in our project. Since
October the Alternate Media Center has been responsible for this program.

Popkin, Ray. Washington Community Video Center. Open
letter concerning survival information, Summer 1972.

At the April Video Conference, there were
several problems common to us all. Mainly that,
because of lack of communication we were having
trouble with distribution, duplicating efforts
in the production of certain kinds of tapes and
we were suffering from a lack of information on
what other folks were doing in our fields of interest. . . . We are now producing a manual on
the use of Video Tape for Survival purposes.
That is the use of video in the fields of health,
nutrition, legal aid, housing, food, etc. We
hope this manual will be of use to Video groups
but also to community groups, colleges, libraries,
clinics and alternative institutions. Not only
do we hope to list what tapes are available in
this field but how they can be used, how to get
equipment to show them and how tapes can be made.
We hope that people will send reports on their
projects which we can reprint so that there will
be an exchange of ideas, mistakes, successes,
etc. We also hope when people send us a list of
their tapes they will describe the process in
which it was made and some of the uses they fore-
see for the tape.

The Videocassette & CATV Newsletter, vol. 2 (September 1972).

During the recent Republican and Democratic
political conventions held in Miami Beach, a group
of blue-jeaned, long-haired "videofreex" produced
a 60-minute program on each of the two sessions.
Mr. Shamberg--the author of "Guerrilla Television," stated that the programs were aimed at
cable because it allowed the group to program on
a full-time basis and to define what CATV should
be. The group all share a passion for and a
faith in the future of half-inch tape as a medium
for electronic communication.

_____. A subscription brochure, September 1972.

Nothing has come on the entertainment, educational and technological scene during the last few years with as much force and excitement as the videocassette concept.

1973 will also be the year when, after many promises, the wonders of cable tv are expected to emerge into the marketplace.

These two aspects of the electronic revolution in communication--videocassettes and cable --will create a multi-million dollar business before the end of the decade. Thousands of people, and hundreds of new companies will participate in the growth of these vital new media of the future.

VidNews, vol. 2 (September 18, 1972).

Walter Dale, Video Project Director, Port Washington (NY) Library manages a project based on the use of two 1/2-inch camera/VTR porta-pak combinations plus two stationary decks used to collect community event tapes and make them available as "localcasting."

The project was funded by the NY Council of the Arts to demonstrate the ability of a community to use video to mirror its own activities. Dale said his people-to-people video has been made available to 8,000 community residentson an in-the-home basis. The porta-paks are issued to anyone who wants to use them for community activity recording.

Washington Community Video Center. "Cable Television in the District, A Community Strategy Session." Program announcement, June 29, 1972.

The Community Video Center is a video-cable project which was until recently associated with the Division of Community Education, Federal City College, and is now an independent research-action-production-education group in the city of Washington. The Center's objective is to prepare the community for cable television and to involve citizens in the decision-making process before and after cable comes to the city. We are interested in using videotape to solve social problems and to train residents of the city for new jobs. The Center has been financially supported by the Eugene E. Meyer Foundation.

Reports and Papers

Association for Educational Communications. Position Paper
on Community Antenna Television, March 1971.

> Urge your mayor to consider the possibility
> of municipal government building the local CATV
> system and leasing it back to a local, nonprofit
> organization or to commercial groups. Urge your
> board of education to seriously consider the pos-
> sibility of applying for a franchise to operate
> the local CATV system. Thoroughly explore with
> local civic leaders the possibility of creating
> a special, nonprofit corporation, perhaps in co-
> operation with your local noncommercial radio and
> television stations, to apply for a franchise to
> operate a local CATV system.

Center for the Analysis of Public Issues. Public Access
Channels: The New York Experience. Report for the Fund
for the City of New York, March 1972.

> The regulatory problem facing the local
> franchising authority is an exceedingly subtle
> one, if carried out properly. Conventional regu-
> latory problems--the quality of the signal, the
> subscription price, etc.--might easily be left
> to market forces in the case of a fledgling indus-
> try such as cable television. The City does need
> to pay attention, however, to the kind of equip-
> ment being installed to ensure that it will be
> compatible with the technological advances most
> likely to occur in at least the relatively near
> future, and should also take an aggressive and
> positive role in developing experiments in pub-
> lic-access and municipal usage. It must also be-
> come much more involved than it has to date in
> problems of compatibility of equipment and sche-
> duling inconsistencies between the companies,
> which on several occasions have presented frus-
> trating problems to would-be users.

_____. Public Issues. New York CATV supplement no. 2,
August 1971.

> "The kind of information sent out via cable
> television during the next two or three years may
> determine what will be sent out for the following
> twenty," says George Stoney, director of the Al-
> ternate Media Center at New York University.

119

The Center, which is funded by a two-year grant from the Market Foundation, has modest facilities for instruction and for production of videotape materials.

Stoney draws extensively from his experience at the National Film Board of Canada, where he directed the community-based film project, Challenge for Change. In that project a small crew, including film professionals and a community organizer, put their services at the disposal of a succession of remote Canadian towns to develop local, self-sustaining community dialogue. Using videotape equipment, the Alternate Media Center will expand the range of these experiments in the United States.

_____. Public Issues. Supplement no. 1, July 1971.

Where the revolution's really at . . . is with the proliferating Porta-Paks. With a portable video-tape recorder, you can go almost anyplace and record almost anything--which explains why such units are the favorites of underground videotape makers eager to present viewpoints seldom seen on commercial television.

Why tape and not film? There are a multitude of reasons, and all of them imply, as one critic claims, that film is yesterday's technology and may well be superseded by tape, whose advantages are manifold.

[. . .]

Free access public TV channels have the potential to revolutionize the communication patterns of service organizations, consumer groups, and political parties, and could provide an entirely new forum for neighborhood dialogue and artistic expression.

Committee to Study Cable Television in Monroe County, New York. Cable Television in Monroe County. Final report, December 1971.

The cable system that Cable Television of Rochester proposes would have a central distribution center and six area centers--subject to separate programming. A dual cable would be laid which would have an eventual theoretical capabil-

ity of fifty channels on each cable or one hundred channels for the system. The actual capacity by the end of 1972 is more likely to be in
the neighborhood of thirty channels each or sixty
channels in all. There would be a two-way capability consisting of a digital return to the com-
puter. There would be seven or eight origination
centers with complete studios. Material filmed
[sic] in those origination centers would be fed
to the central distribution center and then trans-
mitted throughout the system. The intent is to
wire the city.

One of the basic points made was that they
would attempt to provide programming in the public interest and access channels although their
franchise does not require them to do so.

Changing Communications. A National College Conference on
Cable Television sponsored by National Cable Television
Association at George Washington University, Washington,
D.C., February 11, 1972. A response by Sol Schildhause,
chief, CATV Bureau, FCC, to a question concerning agency
counselling of local communities:

Franchising, of course, is a local matter, as
you know, and we mean to leave it that way, so
far as we can tell, for the near term, and it
makes some sense, because there are probably
ten thousand places in this country where there
are going to be cable franchises awarded and it's
very hard from Washington, D.C. to know what the
local requirements and what the local demands and
needs are.

We are available on call all of the time and
we have for years been getting calls from local
jurisdictions. I want to say what I think you
probably already know, that there is a shocking
lack of information about cable out in the field.
A lot of those people are awarding franchises on
the basis of very, very little information about
this business.

Dordick, Herbert S., and Jack Lyle. Access by Local Political Candidates to Cable Television: A Report of an Experiment, November 1971.

Unlike over-the-air broadcast television with
its relatively few channels, cable television

(CATV) has enough channels to carry a wide range
of programming to meet local community needs as
well as programs of national significance. One
attractive prospect for public service is in providing access to the medium at no cost to local
political candidates. One such experiment, with
the opportunity to watch, measure, and evaluate,
was made available to the authors. This experiment was carried out during the November 1970
elections in the community of Waianae on the island of Oahu in Hawaii. It provided an opportunity to obtain some insights into the problems
and prospects of using cable for locally originated political programming.

Feldman, N. E. Cable Television: Opportunities and Problems in Local Program Origination. A report prepared for
the Ford Foundation, September 1970.

The purpose of this report is to [examine and
evaluate] past CATV experience with local origination in three quite different settings: (1)
Canada, particularly with respect to the two CATV
systems in Montreal, which are among the largest
in the world; (2) Dale City, Virginia (a relatively isolated suburb of Washington, D.C.), with
grass-roots television originated by community
groups over a small cable system serving a single
tract of homes; and (3) Lakewood, Ohio (a nearby
suburb of Cleveland), with a small CATV system
originating a wide variety of material.

Lamb, Edward. Tomorrow's Communications. Conference on
Communications, Center for the Study of Democratic Institutions, Santa Barbara, Calif., May 22, 1972.

We can shout, and rather successfully, for
meaningful regulations with guarantees for cost-
free public access. The problems of regulation
are being passed upon by technically unsophisti-
cated city and state officials. Lobbies are effective in such a climate. Even if cable grows
into a common carrier status, and I doubt that
this will occur, declarations of intent or regulations are not necessarily a guarantee of public
access or public participation. These channels
are going to be owned by the most potent economic
and political forces in the United States--and

122

the history of our other utilities indicate that
the public interest may be a slogan and private
profit and result. If we talk about the future
of cable TV in the United States, I, for one, prefer to be realistic about the social consciousness
of private industry in a "free enterprise" economy.

New Samaritan Corporation, North Haven, Conn. Press release, March 27, 1972.

> The cable television system proposed for the
Waterbury area by the New Samaritan Corporation
of the Connecticut Conference of the United Church
of Christ will be built and operated "by Waterburians for Waterburians," a spokesman said here
today.
> The New Samaritan Corporation, headed by the
Rev. Arthur E. Higgins, North Haven, was organized
in 1970 by the Connecticut Conference of the United
Church of Christ to "take action directed toward
the improvement of the conditions of people in the
community." In its two years of operation, it has
undertaken a number of government and foundation-
funded community service programs, including the
construction of several low-income housing projects
for elderly and minority people, three of which are
in Waterbury.

Peters, Robert W. Cable Communications Revolution. A
paper presented to the First International Cable Television Market Conference, Cannes, France, March 6, 1972.

> The industry as a whole . . . has not pur-
sued the development of local origination aggres-
sively for several reasons. First, most of the
scarce resources of this industry have been devoted to constructing new cable distribution facilities, acquiring additional franchises, and at-
taining more favorable federal regulations. Second, there has been considerable uncertainly about
CATV regulation, specifically regulation pertain-
ing to local programming. Third, there was con-
fusion about selection, cost, and operation of
video production equipment resulting from a continuous stream of new, lower cost products. Perhaps most important, however, was the pervasive
sentiment that local origination was supposed to
be like broadcast television. . . . Local origi-

nation might best be described as "people to people" communications or "neighborhood broadcasting."

_____. The Growth Prospects for CATV. A paper presented to the Fourth National Meeting, Information Industries Association, New York City, April 12, 1972.

With an increasingly hazy crystal ball, the third phase of the development of this new technology, namely changing life styles, can be envisioned for the middle to late 1980s. The two most pervasive impacts will be the substitution of communications for transportation and the increased use of the video medium in place of the printed medium. Historically, people have had no choice but to travel to interact and to read to be informed. Inactive broadband communications greatly expands the choices available to the individual. Like any new technology, however, the advent of broadband cable communications also possesses some real threats such as the invasion of privacy and the prospect of a society of urbanized and suburbanized hermits.

Publi-Cable, Inc. Position paper, November 1971.

PUBLI-CABLE, INC., is a consortium of individuals representative of various educational, public service, voluntary, and community groups concerned with cable communications, particularly its non-commercial possibilities. Cable communications have the potential for allowing the full richness and variety of cultural life in the United States to be expressed in mass communications. . . . To assure that this potential of CATV is reached, however, it is necessary that representatives of public-interest organizations of all kinds and the public work together, both to educate themselves, their organizations, and the public concerning the capabilities of CATV, and to attempt to influence policy concerning the establishment and regulation of CATV systems.

Schlafly, Hubert J. The Real World of Technological Evolution in Broadband Communications. A report prepared for the Sloan Commission on Cable Communications, September 1970.

Coaxial cable gives us a multiplying factor
for the electromagnetic spectrum. If the incen-
tive is great enough to "wire a city" we can re-
produce our present communications capacity each
time we choose to install a coaxial cable system
for distribution of signal to the homes in that
city. Each such cable system represents a quan-
tum jump in our communication capability. Instead of husbanding the assignment of radio channels for a limited few, the economy of scarcity;
we can now challenge the imagination and energy
of programmers, service suppliers, community psy-
chologists and educators for productive utilization of the communication channels that can be
made available.

_____. Your Personal Genie in the Cable. A paper pre-
sented at the Whitehouse Conference on Aging, Washington,
D.C., November 30, 1971.

The important fact is that you have become
a PARTICIPANT in a Communications System and you
are not just a powerless recipient. You can take
part in community affairs, express your opinions
and cast your vote. Television is no longer a
spectator sport where you must watch what some
person on Madison Avenue deemed to have "mass appeal."

Schwartz, Louis, and Robert A. Woods. "A Marriage Proposal: Cable Television and Local Public Power." Public
Power, November-December 1971.

Whether for rural or for urban settings, the
cable industry faces not only an attractive present, but an even more promising future, as its
potentialities become every-day realities. Despite its high capital and operating costs, and
its checkered regulatory history, it represents
the most challenging, as well as the most inviting, opportunities for investigation and investment by any group, public or private, desirous of
applying and extending telecommunications for both
public-service and private uses.
For local publicly owned electrical utilities,
CATV offers a new and natural extension of the services they now provide.

Singer, Arthur L. <u>Issues for Study in Cable Communications</u>. An occasional paper from the Alfred P. Sloan Foundation, September 1970.

> A cable television system, once sufficiently large and sufficiently interconnected, can provide two quite new kinds of network. First, by linking selected neighborhoods wherever they can be found, it can create (for example) ethnic networks. Second, by linking selected (or self-selected) residential units independent of their geographic location--and because of its copiousness, the system in the end will be able to do exactly that--it can create networks responsive to particular interests: networks, for example, of those who customarily read <u>The New Yorker</u>, or <u>Harper's</u>, or <u>Foreign Affairs</u>, or networks of practicing physicians.

Society of Motion Picture and Television Engineers. <u>Synopses of Papers; 111th Technical Conference</u>, New York City, April 30-May 5, 1972.

> To make public access channels work, production facilities as well as technical and programming help is required. Since its inception, public broadcasters have been producing community oriented programming on shoe-string budgets and are thus uniquely suited to provide these services to groups desiring to use the public access channels. In some communities it may be in the best interest of the CATV operator and the public broadcasting organization that the public access channels be programmed by the public broadcaster and the CATV operator simply supplying the channel.

TelePrompTer Corporation. <u>First Annual Report on Program Origination</u>, submitted to the FCC, Washington, D.C., September 1971.

> Buying new equipment and adding employees are not enough to assure creative, responsive, high quality programming. However, care has been taken to see that system program managers are always fully attuned to TelePrompTer's philosophy of community-oriented programming, as well as being technically proficient.

126

Yin, Robert K. "Cable Systems and the Social Geography of Dayton." Cable Communications in the Dayton Valley: Basic Report, January 1972. L. L. Johnson, W. S. Baer et al. A report prepared by Rand with financial support from the Charles F. Kettering and Ford foundations.

If local programming among districts is to vary, we need to inquire into whether the configuration of cable districts should influence, or be influenced by, the geographic pattern of population differences. Such a quest for information is fraught with questions involving value judgements. For example,

-- What constitutes a population "difference"?
-- Who is to make final judgements, and what data are to be used?
-- Given that different geographic patterns can even be defined, should cable-TV systems be designed to reinforce existing differences or attempt to override them?
-- Should cable-TV systems accommodate current or anticipated patterns of social geography?

_____. Cable Communications in the Dayton Valley: Summary Report.

A troubling aspect frequently mentioned in past discussions is that many potentially attractive services would be oriented toward low-income groups; yet many low-income people may not be able to afford cable at existing rates. Some observers feel that as a national policy, cable services should be extended to most or all households. As a case in point, the Citizens Advisory Committee on Religion--one of the committees that has met periodically during the course of the RAND study --recommends that "some means be devised whereby this system is made operational in every home regardless of the ability to pay and that program time be available to every citizen on a first come-first served basis." One way to reach this goal is by generating large enough revenues for lease services to enable the fee for ordinary home subscribers to be reduced to near zero. The day of near zero subscriber fees is far off, and one would be on slippery ground at this time to attempt financial projections on that basis.

Yet the objective of generating increasingly
large revenues from leased channels, to the benefit of other users including home subscribers,
is a worthy one.

Personal Letters

American Association of School Administrators. The School
Administrator. Letter from editor Martha A. Gable, June
1972.

      Paul B. Salmon, executive secretary, AASA,
has on a number of occasions urged superintendents
to exercise leadership in their respective communities to make sure franchises protect public service interests. A discussion group at our recent
annual convention in Atlantic City focused on
"What Administrators Should Know About Cable Television."

Broadcasting and Film Commission, National Council of
Churches. Letter from S. Franklin Mack, supervisor, Cable
Information Service, June 29, 1972.

      It happens that I have interviewed key person-
nel in both Alternate Media and Open Channel
within the past week. Alternate Media's Mrs. Red
Burns and her colleagues are all over town stir-
ring up block associations and community agencies
to become actively interested in cable. The day
I was there word came in that the Washington
Heights Association had voted $25,000 to set up,
equip and operate a cable center there, with Alternate Media's assistance. Phyllis King [sic]
. . . talked about her experience helping people
get organized and running in their own momentum
over a three month period in Reading, Pa., and
one of the other staffers talked about the project they've gotten off the ground in Cape May,
N.J., and for which they'd just received a $5,000
TelePrompTer grant. They're what some call "video
people." They don't tell people how or what to do,
but let them learn by doing. But they do get folk
involved in the use of equipment and the preparation and cablecasting of programs of a grass-roots
nature. It's a matter of helping people to help
themselves, and involves a good deal of workshop
provision.

Washington Community Video Center, Washington, D.C. Letter from Nick DeMartino, July 24, 1972.

We are at the stage, in the District of Columbia, where there is not even an ordinance governing cable television at this point, much less any public access channels to be utilized. We are using videotape extensively--in several community projects and with many community organizations that ask us for assistance. However, we think it is important to involve the community in the process of determining what kind of structure cable will have in the city before it is developed. Ordinarily, if the community is involved at all, it is only when the ownership and organizational structures have been established. Consequently, we are organizing in D.C. around the principle of community participation in the decision-making process.

West Virginia Community Television Association. Letter from Vice President William G. Thompson, May 19, 1972.

Until the libelity problem is settled, our state association and its members will be reluctant to establish a so called public access channel. Once this problem is settled, our cable operators and State association will take an active part in public access.

The Press and Periodicals

Broadcasting. "Cable Holds New Rules up to ACLU's Proposals," March 6, 1972.

The National Cable Television Association and 20 system operators last week attacked in two separate filings with the FCC a proposal by the American Civil Liberties Union that essentially would reduce cable to the role of common carrier.

The only thing the FCC did not do in its regulatory package, NCTA said, was to call CATV a Title II common carrier. Instead, it said, the commission has permitted "latitude and experimentation" in cable, and has stated that CATV operators "are neither broadcasters nor common carriers within the meaning of the Communications

129

Act. Rather, cable is a hybrid that requires
identification and regulation as a separate form
of communications.

_____. "Open Access: What Happens?", May 1, 1972.

"Public access is bound to work much better
in small communities," says John Sanfratello, the
program manager for Sterling Manhattan, "because
people often deal on a personal basis with the
mayor, the chief of police, the principal of the
local high school. And everybody knows what his
neighbor is doing: the community involvement is
there to begin with. In New York, people live
behind locked doors; the average New Yorker barely
has a nodding acquaintance with the guy who lives
next door to him."

_____. "Stay Sought for Cable Rules," March 13, 1972.

The present rules place an "unfair burden"
on smaller cable systems since such a system--al-
though it might have only 200 subscribers--must
construct a 20-channel two-way system with four
origination channels if it falls within 35 miles
of a top-100 market.

Cape May Star and Wave. "Community Cable TV Takes on New
Meaning," February 17, 1972.

Now the people are emerging from their commercial isolation and taking active roles because
they can view the fruits of their labor on cable
television.
The citizens of these South Jersey communities are seeking improvements and raising questions and issues which would formerly have gone
unexplored and unresolved, says TelePrompTer.
"Now, through the efforts of two students
with borrowed equipment and the help of the local
cable system, people are waking up to the problems
of their community," TelePrompTer says. "Perhaps
because they now have a means to make their interest public."

Cox, Kenneth A. "Broadcasters as Revolutionaries." Television Quarterly, vol. 6 (Winter 1967).

Broadcasters are expected to serve minority
tastes as well by providing news, agricultural pro-

grams, outlets for local talent, religious programming, programs designed for significant minority groups in their audience, discussions of con-
troversial issues of public importance, educational programs, and editorials. For a variety
of reasons, enforcement of this principle of "public service programming" has never been as vigorous as implied by commission statements.

Deihm, Donald L. "Have Something to Tell Community?
Here's Your Chance: Free TV Time Available to Anyone."
Reading Eagle (Reading, Pa.), January 23, 1972.

> It doesn't matter whether you have an ax to
> grind or someone to praise, you can state your
> case on television. That's right, if you or your
> group has something to say, there's free air time
> waiting for you on Channel 5 of the Berks and
> Suburban TV Cable companies.
> What's more, you can tell it your way. You
> tape the show and edit it yourself.
> Don't worry about not having the technical
> knowledge of a professional producer.
> "No experience is necessary," according to
> Mrs. Phyllis C. Johnson of the Alternate Media
> Center at New York University. Mrs. Johnson has
> just arrived in Reading to coordinate this noncom-
> mercial pilot project.

Evening Outlook (Santa Monica, Calif.). "Cable Television
Lures Churches," June 17, 1972.

> Religious organizations that have campaigned
> for years to liberate church programs from radio
> and television "ghetto hours" are seeking active
> participation in the cable television networks
> springing up across the country.
> One of the leading voices calling for diver-
> sification of cable television is the National
> Council of Churches, whose officials maintain that
> if this new method of disseminating information is
> to realize its full potential, there must be an
> informed and active citizenry.
> To that end, the Council has organized a Cable
> Information Service under the direction of its
> Broadcasting and Film Commission. The service
> is issuing a monthly digest covering all aspects
> of the development of cable TV and is providing

experts to consult with those wishing information
and planning to take action.

Gent, George. "Public Access TV Here Undergoing Growing
Pains." New York Times, October 26, 1971.

Its impact, despite widespread initial publicity and the infectious enthusiasm of its advocates, has been muted by such practical realities
as high production costs, technical problems and
lack of public awareness. Nevertheless, most of
those involved in public access TV, including the
cable operators, city officials and catalyst groups
for community utilization believe the problems are
soluble and the future is limited only by the imagination of those who use the system. . . . "Our biggest problem is informing the public that they can
go on television and say whatever is on their minds.
People are used to thinking of TV as something
someone else does, not as something they do. We
have to overcome that inertia."

Gould, Jack. "Waterbury Officials Planning for First CATV
City." New York Times, January 20, 1972.

The first effort to wire up an entire city
for cable television will be undertaken in Waterbury, Conn., as a non-profit project by a group
of church leaders, businessmen and local citizens.
The contemplated venture could avoid the in-
terminable flow of studies of cable TV and provide
a practical test of how the public responded to
programs specifically designed to meet their local
needs and interests. He predicted it would take
three years to wire up Waterbury, which was specifically chosen because of its ethnic diversity, un-
employment and less than ideal TV reception in
valley conditions.

The Jersey Journal and Jersey Observer. "Hoboken Will Open
Videotape Workshop" (Hoboken, N.J.), June 8, 1972.

A videotape workshop will be offered in Hoboken
this summer to give citizens an understanding of
"Public Access" cable television and the necessary
skills required to produce their own community programming, it was announced today by Mayor Louis De-
Pascale.

The project has the support of Donald Aisenbury, general manager of Cablevision of New Jersey.
"We expect to be transmitting to Hoboken subscribers this fall," he said. "We are pleased that the community is showing interest in using the public access channel which we will provide."

The workshop will be conducted by Miss Chambers; Sam Fiedler, also with the Town Band project; and Eileen Connell of the Alternate Media Center of New York University.

Kaull, James T. "CATV Program Board Suggested as Censor." Providence Journal (Rhode Island), July 24, 1972.

A community programming board might be the proper body to guard against pornography and libel on public-access soapbox channels, a cable television franchise applicant said yesterday.

Richard M. Galkin, president of Rhode Island CATV Corporation, said this censorship task is one he would prefer to relinquish, although the Federal Communications Commission insists that cable operators must bear the responsibility.

Kline, Anna. "Rand Report on a Regional vs. City CATV in Dayton Area Cues Conflicts; Push Interconnected 6-Cable System." Variety, February 16, 1972.

Dayton is the first area of its size to consider the concept of setting up a regional system.

Three spokesmen from minority group organizations were bitter in their reaction to the Dayton cable study, and also criticized the communications media in general, the FCC and the "white power structure" throughout the nation. All said CATV represents the first opportunity for blacks to become involved significantly with mass communications, but that the Rand-Ford-Kettering study virtually ignored blacks. They said the local plan should demand substantial minority ownership.

Korman, Frank. "Innovations in Telecommunications Technology: A Look Ahead." Educational Broadcasting Review, vol. 6 (October 1972).

The traditional classroom and school building may be used primarily for social interaction as instruction is brought to the students' homes. Recent and projected attempts to settle the integration question may become spurious as cable technology obviates these considerations. The social impact of cable will alter our basic economic and political institutions in unforeseen ways.

Korsts, Anda. "Public Access and Public Housing." Cable Report, vol. 1 (April 1972), supplement to Chicago Journalism Review, vol. 5 (May 1972).

Rumor has it--and rumor is the major source of information regarding Chicago's plans for cable television--that one of the main reasons Mayor Daley is taking his time in deciding how and when the city will be wired is that he is apprehensive about the unknowns of public access.

If anyone has the right to go on television on a first-come, first-served basis via public access channels (as the law requires), who knows what will be cablecast in Lawndale or Uptown--to name just two of the city's most depressed neighborhoods. It seems a safe bet that, at the least, the channels would represent a way of developing an independent minority leadership."

_____. "Video Groups Plan Programming." Cable Report, vol. 1, supplement to Chicago Journalism Review, vol. 5 (March 1972).

Chicago can't look to national foundations for help in funding public access experiments. It's too late: The national organizations have already done the work by supporting many initial research and promotional projects--particularly in New York City. They have no interest in repeating themselves and seem to be putting their priorities into cable "think tanks."

Public access research will sink or swim on local funding. And, given the logic of the medium, perhaps that is as it should be.

_____, and Larry Filler. "Two Experiments in Public Access." Chicago Journalism Review, vol. 5 (June 1972).

Cable operators pay a franchise fee to the city. This money should be set aside to be fed back for the support of public access facilities. There should be an entity or a support structure like "Open Channel":--that is devoted exclusively to providing training, facilities and promotion.

The cable operator can't do it. He's concerned
with operating the cable and making it profitable."
Theodora Sklover

The Mercury (San Jose, Calif.), March 18, 1972.

The Santa Cruz Community Council has launched
an effort to establish free public access television, giving all local groups and individuals a
chance to put on their own programs.

In a separate action, the council called on
the Santa Cruz City Council to place the public access initiative petition on the June primary ballot
"in order that the democratic process be affirmed."

The city contends that the initiative proceed-
ing is illegal.

The Community Council, made up of representatives of 36 different local organizations, urged
that the equipment and other expenses involved in
local programming be paid for by franchise fees
paid to the city and county by a Santa Cruz cable
television firm. The revenue now goes into the
general fund of the city and county.

Mitchell, Henry. "Cable TV Talks." Washington Post, February 12, 1972.

Chairman Dean Burch of the Federal Communications Commission told a college conference on cable
television yesterday that he felt it was the federal government's job to make cable television possible and then leave it to the entrepreneurs and
citizens to develop it.

On "public access" the government-approved
scheme by which anybody can use a channel for five
minutes--he said there is a difference between access and audience.

The mere fact that you're on the air doesn't
mean people necessarily will watch.

New York Times. "Public Cable TV Urged in Detroit," May
14, 1972, p. 111.

A committee of private citizens recommended
today after a year-long study that any cable television system built in Detroit should be publicly
owned and controlled.

"Those who control cable systems will, in the
future, control the flow of information, entertain-

ment, news, social and commercial services to the public," the committee said.

"The validity of cable television for Detroit lies not in its potential for a profitable private business, but in this city's need to overcome discrimination in access to communication."

While a public authority would build a cable system for the entire city, private interests could operate the system, the committee said. It suggested that the 142-square mile city be divided into five areas with boards elected by subscribers handling some programming in each of the five areas.

Newsweek. "Do-It-Yourself TV," January 3, 1972.

The first [sic] and still the largest, free-access experiment in the country is a nonprofit organization called Open Channel, which broadcasts public-service programming to some 90,000 cable-TV subscribers in Manhattan. Designed primarily to afford ordinary citizens the chance to express themselves on political, cultural and social issues, Open Channel has given free air time to groups ranging from the Boy Scouts to local supporters of black radical Angela Davis. So far, the project has met with mixed success, but it is being watched closely by government and broadcasting officials across the country as a showcase for the future of do-it-yourself television.

O'Connor, John J. "TV: Public Access Fete." New York Times, July 7, 1972, p. 63.

According to a young woman from the Downtown Community Center, neighborhood interest in public access is surprisingly strong. The only problems are logistical: getting equipment and protecting it from robbery, and getting the company to lay cable lines in the relatively poor area.

Yesterday the neighborhood even managed to get its protest directly on cablevision. Shortly after noon, a Downtown Community News segment was interrupted with the comment that "the people of the Lower East Side have nothing to celebrate . . . for the next 20 minutes, you will be watching what the people of the Lower East Side have in the way of cable TV--that is, nothing."

And for 20 minutes, the viewer could watch a
blank screen. It is that kind of celebration, a
centralized honing of a tool for electronic decentralization. It is unusual. And it is undoubtedly
significant for the future of cable TV.

_____. "TV: Added Exposure for Public Access on Cable."
New York Times, June 13, 1972, p. 87.

Few areas of communications are less understood,
or even generally less known, than that of "public
access" on cable television.
The public access experiment began in New York
in July of last year as part of a city requirement,
in return for the privilege of ripping up streets
to install cable lines, that Sterling and TeleprompTer each set aside two channels that would
be accessible, without charge, to the direct participation of the public.
On a programming matter, there is little or
no evidence that, as was feared in some quarters,
public access has been dominated by freaky or off-
beat groups aware of the technology of videotape.
Of all the programs submitted for public access,
only two have been rejected as unsuitable, both
for varying degrees of sexual explicitness.

_____. "Public Access Experiments on Cable TV Advanc-
ing." New York Times, June 6, 1972, p. 82.

Mrs. [Red] Burns . . . says she has lessened
her "hangup" over attracting audiences. The important thing, she insists, is to develop the skills
and technical integrity that will enable the potential of public access to be realized, and the audience will grow with that development.
In New York an enlarged access operation is
slated at the Alternate Media Center with the help
of $10,000 in equipment being donated by Sterling.
Mrs. Burns notes that the public is only beginning
to be aware that something called public access
even exists. Interest, she says, is growing rapidly.
Eventually, it seems, television's monologue
may have to make room for cablevision's dialogue.

Oppenheim, Jerrold N. "City-Owned Cable Systems." Cable
Report, vol. 1 (June 1972), supplement to Chicago Journalism Review, vol. 5 (July 1972).

Telephone systems are common carriers: everybody has an equal shot at getting a telephone line
on a first-come, first-served basis and the system
is obligated to maintain a number of telephone
lines a year or so ahead of demand so that there
is almost never a substantial wait for a telephone
line.

If cable television were organized along similar common carrier principles, censorship on munici-
pally-owned systems would be impossible. Where it
is not so organized, municipal ownership may bring
more trouble than it is worth.

_____. "Industry Huddles Here." Cable Report, vol. 1
(July 1972), supplement to Chicago Journalism Review, vol.
5 (July 1972).

Cable is a regulated industry, though one
could hardly guess that from the pandemic confu-
sion of cable legislation. Federal Communications
Commission Chairman Dean Burch was refreshingly
candid: "I'm frank to confess that the [FCC's new
cable] rules are terribly complicated. Cable begins its new era in a regulatory maze. . . . At
least we all begin life equal: we're as perplexed
as you are." But Burch's way out of the maze was
typical FCC lawlessness: "We'll permit special
showings and grant special exceptions from the
start. . . . I want to stress that the rulebook
is not carved in stone. . . . It is open to change
and refinement." In smoke-filled rooms, no doubt.

Phillips, McCandlish. "TV of the People Operating on
Cable." New York Times, September 25, 1971.

In places as dissimilar as Tullahoma, Tenn.,
and lower Manhattan, citizens are turning a video
eye on their own communities and televising the
results to their neighbors, without professional
intervention.

In Cape May, N.J., and Charleston, W. Va.,
people who once regarded televising as inaccessi-
ble--are finding that it is not a great deal harder
to use than the telephone system.

Citizens in the places named above have been
receiving help from a small office at New York University called the Alternate Media Center. It is
devoted exclusively to cultivating cable television

as an outlet and resource for local, non-profes-
sional communicators.

Playboy. "Shoot & Show," May 1972.

In considering how half-inch video tape and
do-it-yourself portable units may remake society,
some of video tape's more farout theoreticians have
rather interesting ideas. Philip Morton, a young
assistant professor in experimental video at the
School of the Art Institute of Chicago, insists
that video tape is not product but process (when
not recording, the video camera shows your image
on the monitor scope simultaneously--in what Morton calls "no-time"--but from a completely different angle, which is oddly upsetting; what's happening is not two different actions but a single one,
in which the image feeds back to the performer and
vice versa), and that instant feedback will subtly
but inevitably alter the behavior, and perhaps
even the nature, of whoever is watching. Morton
believes that the identity crises so familiar to
today's generation may never occur at all to a
generation that's used to having itself fed back
as information at a very early age.

Reisig, Robin. "The Electronic Soapbox: 'This Is Your
Channel.'" Village Voice, July 8, 1971.

"We spent 10 hours last night taking out all
the dirty words--there must have been 100 fucks
and cocksuckers--and now they want us to put them
all back in," muttered the technician. "It is
pure goddamned prejudice!" shouted the videotape
maker because the last six minutes had, for technical reasons, just been cut out of his videotape
on the Italian-American Civil Rights League. "You
wouldn't cut somebody else! You cut an Italian,
goddamnit! I can't help boiling in my blood! You
fucking liberals!" A girl tranquilly translated
it all into sign language for the deaf. The Sterling television screen went blank and silent because the videotape maker had stormed the broadcast studio just as the announcer was saying, "This
is your channel and you can say anything you want
on it."
It was the first day of people's television
in New York or, cable officials claimed, the coun-

try. Last Thursday Manhattan's public access channels opened to the public. With its enormous channel capacity and low cost, cable television is supposed to provide an opportunity for diversity of
ideas and opinion, for intellectual and ethnic and
community programming, not usually heard on television. The first day would not have been broadcast
on commercial television.

Sonnenfeld, David. "Failure to Communicate." Business
Today, Spring 1972.

No real communication can occur between those
who sell a product to make a profit or a payoff and
those who have been coerced into buying the product. Not until there is popular control (popular
not meaning stockholder) over the exploitations of
business and technology, can there be meaningful
communication and discussion concerning directions
and goals for business and society.
When AT&T stops making guidance systems for
missiles, when Standard Oil stops exploiting the
Third World, when Bank of America gets out of agri-
business and real estate, when the U.S. government
no longer supports and participates in racist global
colonialism, then perhaps there can be meaningful
discussion of what business and the rest of our
country can do together.

University Daily Kansas. "Local Cable TV Offers Public
Access Program," April 18, 1972.

Public access to Sunflower Cablevision . . .
has been available since its first month of operation in January, according to Michael Pandzik,
Cablevision's production manager. "Public access
means we offer production facilities, personnel
and time in order that any group in Lawrence can
present its philosophies and activities on channel
6." Pandzik said the half-hour public access shows,
aired at 5:30 p.m. Thursday and 9 p.m. Friday, cost
about $50 to produce. . . . A public non-profit
corporation would be the answer to some current
problems of public access, Pandzik said. Such a
corporation would be responsible for the liability,
editorship, and scheduling of public access programming.

140

An individual can request air time under the
public access regulations, Pandzik said, but he
preferred to start with the legitimate, organized
groups in the Lawrence area, which number about
250.

Weaver, Warren, Jr. "Cable TV Getting Political Test in
New Hampshire." New York *Times*, February 29, 1972.

Cable operators believe they are functioning
under excessive government regulations, and they
are anxious to recruit allies in the political
leadership of both parties.

Other Sources

Alternate Media Center at New York University School of
the Arts. "The Greenwich Village Charrette" (Catalogue),
Summer 1972.

A Charrette is an accelerated social planning
process, designed to encourage the widest possible
community participation. Coverage of this Charrette
in Greenwich Village on half-inch videotape was the
first project undertaken by the Alternate Media
Center.

A year later, half-inch is taken for granted
on the cable; the FCC has taken a strong stand
against censorship of public access channels by
cable operators; the use of half-inch by nonpro-
fessionals is growing daily. But, in May of 1971,
the Charrette was a place where many alternate
media questions would be put for the first time,
and most of the conditions were ideal for experimentation.

The only hitch was that, at the time (May,
1971), very little of the P.S. 41 community was
on cable. So, auxiliary viewing places were set
up, in addition to the cable coverage. The same
tapes (over 40 hours worth) that were played over
the cable system on a two-hour delay, were also
played over three monitors outside the school on
the top of a truck. When people inside came out
for a breath of air, they were able to see workshops that they had missed or even see themselves
in action two hours before. The response to this
feedback was consistently enthusiastic.

Federal Register. "Rules and Regulations," vol. 37 (July
14, 1972).

> Although the cable operator will have no con-
> trol over program content on access channels, he
> is charged with proscribing the presentation of
> obscene material. It is suggested that to this
> extent, at least, the operator will, in effect,
> be required to exercise control. To clarify
> this area, we are requested to seek legislation to
> grant immunity to a system operating under our ac-
> cess rule. We, of course, appreciate petitioners'
> concern over the liability issue. We still be-
> lieve, however, that existing case law solves most
> problems in this area.

The Last Whole Earth Catalog. Menlo Park, Calif.: Nowels,
1971.

> We are as gods and might as well get good at
> it. So far remotely done power and glory--as via
> government, big business, formal education, church
> --has succeeded to point where gross defects ob-
> scure actual gains. In response to this dilemma
> and to these gains a realm of intimate, personal
> power is developing--power of the individual to
> conduct his own education, find his own inspira-
> tion, shape his own environment, and share his ad-
> venture with whoever is interested. Tools that
> aid this process are sought and promoted by the
> WHOLE EARTH CATALOG.

Source Catalog: Communications. Chicago: Swallow Press,
1971.

> What you will read is not our analysis of the
> communications revolution, but a self-portrait of
> the groups struggling in it. These groups are the
> basis of the Catalog: their causes, purposes, pro-
> grams and resources. From letters they've written
> to us and from phone conversations, we've tried to
> pass along their self-descriptions and to follow
> up on other projects they've turned us on to.

Tarshis, Morris. "Rules Governing Access to Public Chan-
nels." Public Access Channels: The New York Experience,
A report for the City of New York by the Center for the
Analysis of Public Issues, New York, March 1972, Appendix
I.

For the purpose of gaining such experience
and in order to encourage differing uses of the
Public Channels the two Public Channels shall be
governed by different concepts. On one Public
Channel, denominated Channel C in the franchise,
there shall be an opportunity to reserve a particular time period each week, or several time periods each week, in order to permit the user to
build an audience on a regular basis. On the
other Public Channel, denominated Channel D in
the franchise, there shall be no multiple time
reservations, in order to permit a user with a
single program and users with relatively last-
minute requirements access to prime time periods.

United Church of Christ, Office of Communication. A Short
Course in Cable. New York: United Church of Christ, November 1971.

Whether a non-profit community corporation
or a private operator gets the franchise, the cable organizing group should press for the establishment of an officially-recognized independent watchdog agency to protect the public's interest.
The agency should work with the cable to see
that his promises of facilities and programming
are carried out, and that the public has access
to the system on a non-discriminatory basis. This
independent agency could be established under the
cable franchise. Funds to operate it could come
from a fixed assessment of from 2 to 5 per cent on
the cable operator's gross income. Such an agency
should represent the widest possible cross-section
of community composition and interests.

Antioch College, Baltimore-Washington Campus. Videoball
Brochure, 1972.

We think people speak best for themselves,
and that developing skills to communicate their
own information is an important step in gaining
and exercising control over their own lives. 1/2"
video is the tool that makes this process possible
and we're trying to make it a reality by sharing
our explorations and discoveries with Baltimore
and you.
We are part of the new Antioch College, Baltimore-Washington campus. We started as a documen-

tary arts program, but have branched out with VTR
to become an experimental communications support
system for the courses and concerns of both the
Arts and the Social Research and Action centers.
These centers are project-oriented, community
based and the faculty serve as consultants to both
students and community groups. Students can earn
credit toward a Bachelor of Arts degree for involvement in work and independent study projects, life
experience, as well as for course studies.

Watts Communication Bureau. "An Application to the City
of Los Angeles for a Franchise to Install and Operate a
Community Antenna Television System in the Los Angeles
Basin," 1972.

The Watts Bureau has many service objectives,
but principally it will accommodate the Black and
Mexican American residents of South and East Los
Angeles in their efforts to breach the barriers
to communication that exist between and within
their communities and other communities lying
within the Southern California region.

PRIMARY ENGLISH CANADIAN SOURCES

Blue, Art. "This Is Your Road: Bring People Back to
Meaning." Access (Challenge for Change/Societe Nouvelle),
vol. 1 (Summer 1972).

Values relate to a person's identity, how one
sees himself. Well, there are three basic portions, as I see it, relating to identity.
One is cultural, and in order for one to relate to the cultural identity, he must have some
feel for his historical roots running back into
infinity. We see this in such things as language,
myth and religion.
Also, it is important for a person to have a
geographical sense, some way of centering as a person: "I'm from Quebec" or "I'm from Labrador."
He is saying, "That is my community; if you look
there, you will make meaning out of how I am, and
that's important."
Finally, identity has to do with self-esteem
--that is, how positively or negatively he relates
to himself. I suspect that the greatest single

144

problem we have in helping people is to help them
gain a feeling within themselves that stands up
and says, "What I am is worth looking at." Maybe
that is what we are all about.

Breitrose, Henry. "Film Power." Newsletter Challenge for
Change, vol. 1 (Fall 1968).

This concept radically re-defines the role of
the film-maker in ways that transcend the usual ideas
of personal artistic expression or the banalities
of producing grist for a sponsor. He becomes a
bit more than a film-maker, perhaps even a new kind
of social engineer. His responsibilities go far
beyond craft and imagination, and in some awesome
ways extend to all of the results of his actions.
As John Grierson put it in the earliest days of
documentary film, its rationale is "this idea of
social use."

Canadian Cable Television Association. "Local Cablecast-
ing--A New Balance." A talk by Pierre Juneau, chairman,
Canadian Radio-Television Commission (CRTC), delivered at
the Canadian Cable Television Association's 15th Annual
Convention and Trade Show, July 6, 1972.

Recently, in the respected pages of the Paris
daily newspaper, Le Monde, Canada was described as
developing new forms of direct democracy in its
use of cable television. I had been led to believe
by a great number of learned articles, most of them
from the United States, that we had to wait for the
wired city and universal two-way cable technology
for the realization of such Utopian dreams.
Whatever the dividing line between fantasy and
fact in the continuing promotion of the wonders of
cable, it remains a fact, as I've said in other circumstances, that Canada is one of the most interesting mass communications laboratories in the world.
Canadian broadcasting, because it is developing
rapidly, does have enormous problems to resolve
but we do enjoy an abundance of communications
technologies, an abundance of public, private and
foreign broadcasting sources, and we have shown a
pioneering willingness to adapt new technologies
to the needs and conditions of our geography and
culture.

_____. "What Cable Television Really Means to Canada."
An informational release, 1972.

Today, some five million Canadians receive
their television signals via cable. The national
average of cable penetration is approximately 30%,
which compares favourably with 9% in the U.S.A.
Some cities, such as Vancouver, Victoria and London, Ontario, have over 85% of the population con-
nected to the cable television distribution systems.
In a number of cases, therefore, the concepts of
the wired city are close to realization. Cable
distribution systems in Canada soon will pass 80%
of the urban residences in this country.

Canadian Radio-Television Commission. Cable Television in
Canada. A Report, January 1971.

It has been pointed out that there are two
kinds of community programs: the kind in which the
program's producer describes the community, and the
kind in which the community describes itself. Both
are valuable, but in conventional broadcasting, the
second kind is rare.
Most systems that have begun community programming are producing, on the average, about 10 hours
a week. They report interest and excitement from
their subscribers, although it is difficult to estimate just how large the audiences are for community programs. One system, which went to unusual
lengths to ascertain its audience for such programs,
concluded that about 24 per cent of its subscribers
watched them.

_____. "The Integration of Cable Television in the Ca-
nadian Broadcasting System." Statement dated February 26,
1971, in preparation for the public hearing held April 26,
1971, in Montreal.

The Commission is most interested in discussing
the methods by which the community can obtain "reasonable and balanced opportunity" to express its
concerns and nourish its identity through the community channel.
If substantial amounts of community programming are to be undertaken by groups, rather than
directly by the cable system licensee, it has been

suggested that such groups be licensed in some way. The form of licensing, the legal responsibility of the group and its representative character would obviously require further study.

The Commission has not indicated so far that it expects every cable system to engage in community programming, and many have not.

_____. Canadian Broadcasting "A Single System." A report, July 16, 1971.

In the opinion of the Commission, there is an obvious danger that the development and even the policy of broadcasting be determined by the natural tendency of hardware, tools, and machines to proliferate as a result of technological and marketing pressures. Such a proliferation can only occur if the hardware is fed with inexpensive contents. This kind of development leads to wider and wider circulation of programmes without a corresponding increase in the production of messages. Messages from larger centres are spread more and more distantly. This results inevitably in a stretching process, a "more of the same" process where, in the long run, choice is reduced rather than increased and where the medium is indeed the message.

_____. "The State of Local Programming." A paper developed by the Broadcast Programmes Branch containing information obtained from preliminary analysis of 82 questionnaires returned by cable television operators, Summer 1972; Interim Summary, July 1972.

It is clear from these statistics that community programming involving initiation of programmes by members of the community is a viable local channel form. In general terms, and excluding 9 responses from cable systems which have had trouble with the questionnaire, it appears that some 17 systems undertake local origination and/or other programming only, while 54 undertake some amount of community programming in addition to or instead of the other types.

Of these systems in this last group all 54 did at least 2 hours of community programming during the week. 40 produced between 2 and 5 hours and 25 produced over 5 hours during the week.

147

Forster, C. C., general manager, Terracom Cable TV, Mississauga, Ontario, Canada. A letter, 1972.

A very little can be said about our style except that we operate our channel 10 as a community expression vehicle and we are pleased to report to anyone interested that after three years our acceptance has grown to very significant proportions. We have only two staff members involved with the co-ordination of more than twenty hours per week of community programming. The balance of manpower is made up exclusively by community minded and community television interested people from all walks of life throughout our neighborhood.

Grierson, John. "Memo to Michelle about Decentralizing the Means of Production." Newsletter Challenge for Change/Societe Nouvelle, vol. 1 (Spring 1972).

The NFB and its Challenge for Change program will have a new and different opportunity of becoming both objective about, and representational of local citizens and local affairs, with the arrival of local TV by cable or otherwise. It is probable that government permits to operate these services will be dependent on the community being represented in the production management by a community league or something of the kind. This, as I see it, must mean that municipal authorities, schools, universities, trade unions, industries, chambers of commerce, and other associations, will all have to look to their images and give an account of their stewardship; and, no doubt, all good radicals and true will see to it.

Grose, James, co-ordinator, Canadian Unitarian Council community television programmes. A letter, July 11, 1971.

Community television is just beginning in Canada and stations are seeking programmes from any interested parties. Our aim is to set up production centres in the larger cities of Canada to:
1. Introduce Unitarians and their philosophy to the Community.
2. Circulate tapes around the centres to introduce Candian Unitarians across Canada to each other.
3. To provide tapes to small isolated societies unable to afford Ministers or speakers of note.

To implement this programme, six programmes have been scheduled for production in Toronto this fall, primarily to gain experience. Our findings will then be circulated to other interested production centres. Funds are low so participation is purely voluntary. However, we do have considerable talent available and many of our members are professionals in the field.

Gwyn, Sandra. Cinema as Catalyst. A report on the seminar, Film Videotape and Social Change, organized by the Extension Service, Memorial University, Newfoundland, March 13-24, 1972.

That film could be a catalyst for social change, the instrument of consensus, was first actualized on Fogo Island, Newfoundland, five years ago. Its happening in Canada was not really an accident. From Harold Innis, Marshall McLuhan, the innovations of Expo and Labyrynth, to a world lead in cable television (because, of course, we all like to watch American programs), communications have always been uniquely part of the Canadian experience.

Henaut, Dorothy Todd. "A Few Notes on Regional Projects." Access (Challenge for Change/Societe Nouvelle), vol. 1 (Autumn 1972).

More and more, we are being thrust into the role of counsellors--transmitters of what we have already learned. Most of the letters piling up on our desks have to do with requests for advice, training and encouragement. More people are obtaining access to video, radio and cable; more are having trouble finding effective uses for it. They need help and encouragement in discovering its dynamic as a social tool. They need to discover their own capacities for creative social use of the media. They need to understand that the media are no miracle-makers, that the important element is their human skills, their social commitment.

Communities need help in conceiving and forming media associations. They need pointers on the various possible models, on their advantages and drawbacks. They need help in recognizing the resources of their community, in seeing beyond the static patterns they are used to, in finding allies in unexpected places.

_____. "Powerful Catalyst." Newsletter Challenge for Change Societe Nouvelle, vol. 1 (Winter 1971-72).

> The need for a real exchange of information and ideas among the various groups that make up the fabric of our society grows more pressing every day. Silent poor or silent majority--we are all suffering from the need to influence the decisions that affect our lives.
>
> Instead of being an instrument to facilitate these exchanges, the media, as presently constituted, usually exacerbate these frustrations by filtering citizens' opinions, when solicited through the well-dressed eyes of professional journalists and communicators. Even where there is a certain variety of input into the system (and it is really very slight), the ideas are filtered through the professional forms and deformations until they become predigested product.

_____. "The Media: Powerful Catalyst for Community Change." Educational Technology, July 1971.

> People must become convinced of their problem-solving capacity on a scale that is meaningful to them. This conviction arises as the result of actual experience and participation in a creative social process.
>
> For this reason the means of communication--real two-way communication--must be made accessible to ordinary people for dialogue in meaningful local debate. In this way we would generate a much more vigorous problem-solving capacity based upon local initiative and creativity.
>
> Rediffusion on a broad scale of original and creative solutions, coupled with free information accessible to all, could alter positively the social and environmental situation.

_____, and Bonnie Klein. "In the Hands of Citizens: A Video Report." Newsletter Challenge for Change, vol. 1 (Spring-Summer 1969).

> Video equipment does not create dynamism where none is latent; it does not create action or ideas; these depend on the people who use it. Used responsibly and creatively, it can accelerate perception and understanding, and therefore accelerate action.

The Comite des Citoyens de Saint-Jacques could
have accomplished any of their actions without the
video equipment. We could not say that any time it
made the difference between success and failure.
But it made good things better, and helped people
to grow. It is a useful tool.

MacDonald, Marylu. "Community TV in Canada." Unitarian
Universalist World, vol. 3 (February 15, 1972).

The involvement of Canadian Unitarian Universalists in community television Programming is the
result of the persistence of one man, Canadian Unitarian Council Treasurer, Jim Grose. An accountant
in a provincial government department that makes
extensive use of audio visual aids, he became interested some years ago in the possibilities of
television programming as a means of communication
between the isolated UU groups in Canada. His
gentle pushings and proddings were given new impetus a year ago by the change in broadcasting
regulations requiring cable companies to make one
channel available for community use, and last year's
CUC annual meeting agreed to a budget allocation
of $1000 for the development of videotape programs
for cable use.

McPherson, Hugo. "A Challenge for NFB." Newsletter Challenge for Change, vol. 1 (Spring 1968).

Since its beginnings--through its films and its
unique distribution system--the National Film
Board has been involved in social issues. Challenge for Change is an outgrowth [1966], adapted to
today's conditions, of strongly-rooted Board tra-
ditions. The Board, at its best, is attempting
to make visible social attitudes that are characteristic of Canadians at their best.
What has Challenge for Change accomplished?
Even by the coolest judgements the program has
been astonishingly effective.

Nemtin, Bill. "Our CRTC Brief." Metro Media Print-Out,
vol. 1 (July 1972).

The C.R.T.C.'s reference and viewpoint is
through the broadcast system to the community.
Metro Media tried to introduce the opposite
viewpoint in its brief--from the community to the

broadcast system. We tried to place community
television, in particular, in the context of community information needs. One of our main obser-
vations, here, was the need to organize each programme to reach the intended audience. Our best
programmes have been those with well organized
groups who could communicate with their members.
The other successes in general programming were
accomplished by intensive advertising in other
media. There are many, many communications needs
in an urban community--in the areas of public education, training, group to group communications,
etc. The cable system can be used to fulfill some
of these.

   This relationship with the cable system im-
plies other needs. A facility that can train, ani-
mate, and inform the community in its use of media
is one of these.

_____. "Propaganda and the Creation of Sacrosanct Sys-
tems." Newsletter Challenge for Change, vol. 1 (Winter
1968-69).

   Amateur film-making has increased to such an
extent that one film-maker has suggested that he
is the last generation of professional film-makers.
To some extent the work of the Challenge for Change
program--Fogo Island, at Kwacha House, or with the
Indian Film Crew--can be characterized as decentralization of the power of propaganda.
   Structural change to include a greater number
in the meaningful decisions of a system is a step
towards personal freedom. However, this step can
potentially reinforce existing obstacles to the
perception of meaningful choices. To exercise freedom a person must be able to perceive various
courses in action and have the power to act.

Newsletter Challenge for Change Societe Nouvelle, vol. 1.
Special issue, "Community Cable TV and You," February 1971.

   In Normandin (Quebec), the citizens have
formed a board of directors to administer community
cable. The nerve center of the operation is the
program committee, which receives, studies and
selects projects submitted by members of the community. A team of three people is then selected
to coordinate the production: a production head,

to coordinate everything; a technical head; and
a research and information chief. These lead the
work crews who produce the programs. At the present time there are a dozen crews producing community cable programs, and similar organizations are
being set up in neighborhood towns.

O'Dea, Jim. "VTR and Urban Renewal on Blackhead Road, St.
John's Newfoundland." Access (Challenge for Change/Societe
Nouvelle), vol. 1 (Summer 1972).

The use of VTR in the Blackhead Road area
seems to have been instrumental in helping the people gain the momentum and understanding that they
have reached. I think that this project played a
major role in helping the Blackhead Road community
gain confidence in themselves. It has grown from
a community that just last year had almost given
up to a community that is now trying to influence
the direction of its own affairs.

One of my biggest problems was that, when
the Union began seeing the positive effects of using VTR, they looked at it as the only way to promote change. As a result, I have had to reverse
my position and get them to think of the medium
as only one of many tools they can use. They can
now discriminate better when or when not to use
the medium.

Rogers Cable T.V., Limited, Toronto, Canada, Programming
Department. Introducing the 20th Century Community Centre.
A brochure, 1972.

You'll discover that it's quite a thrill and
a challenge for you and your group to do a programme.

But it's even more important to your community. Because while community programming may be at
a very early stage right now, in the years to come
it's going to develop into the most vital communications source in your community. And it's people
like yourself who can help make community programming the Twentieth Century Community Centre.

Ross, Malcolm. "Small Article in Response to Editorial
Welcome of: The Film as a Poem of Change." Newsletter
Challenge for Change, vol. 1 (Autumn 1970).

The film theatres, as commercially used,
have provided the community with geographically
located collective drop-out zones. The film material networked through the theatres by major film
corporations has to fulfill certain expected economic roles. Therefore most films are strategi-
cally constructed light fantasias or escapist
drama.
Now we have community film projects. Inter-
changeable activities linked with subsequent social
change. This means that the community is now involved in the direct applications of the former
fantasy-factory.
Films, instead of offering a drop-out zone,
could become the focal point in the new communal
awareness, a sort of drop-in, and create toward
a new society.

Ryan, Teya. "Exposing Our Society to Us." Vancouver Sun,
December 30, 1972.

According to co-ordinator Bill Nemtin, Metro
Media has three major focal points, all of which
serve local interests in some way.
-- Service, which includes producing programs
   on public rights, ecology and community concerns, as well as providing video equipment,
   seminars and workshops for the public;
-- Advancing their own community and video
   perception by initiating their own video
   experiments (they are normally contracted
   by various groups to produce programs);
-- And political where they work as "lobbyists"
   by representing what they see as public interests to various agencies and institutions involved with communication.
"We are offering a production service," says
Nemtin, "where groups can get programs that express their opinions made and presented to the public."
Because of limited resources, Nemtin says,
Metro Media gives priority to low income groups
that haven't the chance to make their opinions
known.

Short, R. C. "Why Cable Is Where It's At." ETRAC Newsletter, Ontario edition, Winter 1972-73.

I suggest that the following are some of the
changes taking place in society manifesting current human intentions to identify as individuals:
-- the "average man" is disappearing
-- the tribal mask is being replaced by the
    private face
-- the members of the masses are again becom-
    ing individuals
-- the importance of common denominators is
    yielding to the significance of uniqueness
-- individual knowledge is superseding expe-
    rience in value
-- individual information is the new currency
    of business with credit replacing cash
Cable television is a new technology whose effects synchronize with this new environment of individualism.

Society of Motion Picture and Television Engineers. Synop-
ses of Papers; 110th Technical Conference & Equipment Ex-
hibit, Montreal, October 3-8, 1971.

The media of communication are few: theatre,
print, radio, film, television. And new ones
don't come along very frequently.
    A new one is coming into being: video cas-
settes. In essense they are television in recorded
form. But the new medium is very different from
television even though it uses the same display
device. Video records can be played on demand,
they can be stopped, repeated, served in response
to the viewer's needs.
    The economics are totally different from those
of over-the-air and cable television. This means
for the first time, that the power of television
can be made to serve small, geographically dispersed audiences.

Watson, Patrick. "Challenge for Change." Reprinted from
Artscanada, April 1970.

While the English half of the [National Film]
Board was giving birth to Challenge for Change,
the French unit was developing its own Societe
Nouvelle: and the first film that I saw in this
series was La Petite Bourgogne, a film made with
the residents of a community that planners had decided to "renew" without involving the people who
lived there.

Through the intervention of the NFB crews,
the planners and the residents were brought together, and the development was enabled to take
account of the real interests and needs and de-
sires of the residents. But before the residents
and the planners could meet, the residents themselves had to get together, and the making of the
film was their instrument for doing so.
   Once the dispossessed and the powerless have
access to the means of information they can no
longer be misled by Establishment bullshit.
   And that is in itself a revolution.

PRIMARY FRENCH CANADIAN SOURCES

Landry, Antoine, and Henri Tremblay. La Television Communautaire au Lac Saint-Jean. Etudes et Reserches. Quebec:
Direction Generale de la Planification, July 1972.

   Au prealable, on peut rappeler que notre participation a cette recherche impliquait une affec-
tation a temps partiel seulement. Nous avons ce-
pendent pu compter sur les services d'un etudiant
en sociologie, Paul Deschenes, qui a passe quinze
jours a Normandin pour realiser la plupart des
entrevues aupres des artisans de la TVC. A l'ete
1971 et en janvier 1972, nous avons aussi employe
18 etudiants du niveau CEGEP pour administrer le
questionnaire a Normandin et St-Felicien.
   Nous avons utilise les quatre techniques de
recherche suivantes: l'observation, l'entrevue,
l'entretien et le questionnaire.

Ouellette, Claude, et al. "La television communautaire
de Normandin." Medium Media, 1971.

   La television communautaire est un instrument
que les membres d'une communaute se donnent pour
entrer plus facilement en communication entre eux
afin d'aborder les preoccupations vitales de leur
milieu.
   On pourrait aussi definir la T.V.C. de la
façon suivante: des gens de la communaute produi-
sant des emissions pour les diffuser par cable--
mais cela ne rendraint pas l'idee de ce qu'est la
television communautaire.

La television communautaire, c'est un milieu
qui se projette a l'ecran pour se voir, se parler,
s'analyser, s'animer et mettre en branle les meca-
nismes de changements necessaires.

A FRENCH SOURCE

Guichard, Marie-Therese. "Le cable cassera le monople."
Le Point, December 4, 1972.

Le public doit etre initie, afin de maitriser
les moyens de production. A Carpentras, les tech-
niciens de l'ORTF disposaient d'un materiel simple
et maniable par tous (des magnetoscopes portables
pour reportage individuel) dont ils ont pu apprendre
le maniement aux habitants. Du coup, les habitants
d'Aubignan, un petit village situe a six kilometres
de la, ont realise un face-a-face sur les activites
de la maison de jeunes, jugee trop gauchiste par
les Aubignanais les plus ages. Chaque group d'habi-
tants--les jeunes et les notables--a realise un
film presentant ses arguments. Puis toute la population a ete invitee a assister a un debat. Re-
sultat: les antagonistes ont reussi, en fin de
compte, a trouver un terrain d'entente.

Cette vision idyllique ne prefigure pas obli-
gatoirement l'avenir d'une societe disposant de
la teledistribution. Tous les responsables insis-
tent d'ailleurs sur le caractere limite de l'experience. Elle permettra toutefois aux responsables
syndicaux et politique qui visionneront les films
realises de mesurer les incidences de la television
par cables sur la vie sociale. Et tout le monde
en est certain, c'est d'une veritable revolution
qu'il s'agit. Le moyen d'expression televisuel
pourrait etre demain entre les mains de tous.
Comme si l'on passait du livre unique a la possi-
bilite pour chacun de se faire editer.

GILBERT GILLESPIE is a Canadian who has lived and worked in both Canada and the United States. He has served as producer-director of instructional television at Channel 19 in Kansas City, Missouri.

Dr. Gillespie did undergraduate work at Meisterschaft College in Toronto, the University of Maryland Overseas Program in Germany, Universidad Interamericano, Mexico, the University of Ghana, and California State University at San Jose. He received his M.A. in radio and television broadcasting from the University of California at Los Angeles and, in 1973, his Ph.D. in speech communication and human relations from the University of Kansas.

RELATED TITLES
Published by
Praeger Special Studies

ASPEN NOTEBOOK: Cable and Continuing Education
Richard Adler and Walter S. Baer

CABLE TELEVISION U.S.A.: An Analysis of Government Policy
Martin H. Seiden

CARRASCOLENDAS: Bilingual Education Through Television
Frederick Williams
and Geraldine Van Wart
foreword by Senator Ralph Yarborough

CHILDREN AND THE URBAN ENVIRONMENT: A LEARNING EXPERIENCE:
Evaluation of the WGBH-TV Educational Project
prepared by Marshall Kaplan,
Gans, and Kahn

THE ELECTRONIC BOX OFFICE: Humanities and Arts on the
Cable
Richard Adler and Walter S. Baer

GETTING TO SESAME STREET: Origins of the Children's
Television Workshop
Richard M. Polsky

TELEVISION PROGRAMMING FOR NEWS AND PUBLIC AFFAIRS: A
Quantitative Analysis of Networks and Stations
Frank Wolf